Dynamics CRM Deep Dive: Administration

Mitch Milam

Dynamics CRM Deep Dive: Administration

Published by
xRM Coaches, LLC
6401 W. Eldorado Parkway, Suite 106
McKinney, TX 75070
www.xrmcoaches.com

ISBN: 978-1-938097-04-1 paperback
ISBN: 978-1-938097-05-8 ebook

Editor: Jennifer Franz Milam
Cover: Donna Casey, digitaldonna.com

Acknowledgements

I would like to thank the following people for helping with the design and testing of this material:

Dale Wilken, Kylie Kiser, Ty East, Victoria Doucet, Robert McAnally

Table of Contents

Chapter 1. Create Your Administrative Toolbox ... 1
Chapter 2. Let's Do a Little Housekeeping ... 7
Chapter 3. Error, Warning, and Notification Messages ... 11
Chapter 4. Managing System Jobs ... 25
Chapter 5. System Job Management: Registry and Database Settings ... 33
Chapter 6. Creating System Management Views ... 35
Chapter 7. Automating System Job Cleanup ... 39
Chapter 8. User Management Tips and Tricks ... 45
Chapter 9. Email Management ... 51
Chapter 10. Email Management – Preventative Maintenance ... 59
Chapter 11. Outlook Sync Table Cleanup ... 63
Chapter 12. Monitoring the Windows Event Log ... 65
Chapter 13. Platform Event Tracing Overview ... 79
Chapter 14. Enabling Event Tracing ... 87
Chapter 15. Troubleshooting Using Event Tracing ... 91
Chapter 16. Matching User Error Codes ... 95
Chapter 17. Troubleshooting Development Errors ... 97
Chapter 18. Workflow Best Practices ... 101
Chapter 19. CRM SQL System Jobs ... 107
Chapter 20. CRM Organization Settings Editor ... 113
Chapter 21. Creating a Management Dashboard ... 119
Chapter 22. Email Router Troubleshooting ... 121
Chapter 23. SCOM Management Pack for Dynamics CRM ... 123
Chapter 24. SQL Server Indexing ... 129
Chapter 25. Backups, Backups, Backups ... 133
Chapter 26. CRM Diagnostics Page ... 137
Chapter 27. Problems with Security ... 139
Bonus Material ... 141

Table of Figures

Figure 3-1. Unhandled JavaScript error 12
Figure 3-2. JavaScript error created by a developer 14
Figure 3-3. Generic Dynamics CRM error (platform layer) 14
Figure 3-4. Generic Dynamics CRM error (unknown origin) 15
Figure 3-5. Business process error caused by a custom plugin 16
Figure 3-6. Asynchronous plugin error 16
Figure 3-7. Error sending email from workflow 17
Figure 3-8. Generic SQL Server error 17
Figure 3-9. System Job error detail 17
Figure 3-10. SQL Server timeout error 18
Figure 3-11. Custom plugin error details (configuration) 18
Figure 3-12. Custom plugin error (bad design) 18
Figure 3-13. Custom plugin error (unhandled exception) 18
Figure 3-14. Workflow configuration (child workflows) error 18
Figure 3-15. Asynchronous plugin error (non-CRM related timeout) 19
Figure 3-16. User configuration error 19
Figure 3-17. Custom plugin error (unhandled exception) 19
Figure 3-18. Internal Dynamics CRM workflow engine error 20
Figure 3-19. Users exist without assigned security roles 20
Figure 3-20. Informational message to install the Outlook Client 21
Figure 3-21. Pending Email warning 21
Figure 3-22. System-wide privacy settings 22
Figure 3-23. Personal privacy settings 23
Figure 4-1. System job actions 27
Figure 4-2. System job action failure - multiple records 27
Figure 4-3. Resume System Job 28
Figure 4-4. Failure to resume a system job 28
Figure 4-5. Postpone a system job 29
Figure 4-6. Workflow job retention 29
Figure 4-7. Delete AsyncOperation if StatusCode = Successful 30
Figure 6-1. System Jobs with Messages 36
Figure 6-2. System Jobs with Messages (Recent) 36
Figure 6-3. System Jobs (Canceled or Failed) 37

Figure 7-1. Bulk delete successful workflows ... 43
Figure 7-2. Bulk delete canceled and failed system jobs ... 44
Figure 7-3. Bulk delete system jobs with messages ... 44
Figure 8-1. Client Access License Information ... 46
Figure 8-2. Users exist without assigned security roles ... 46
Figure 8-3. Personal Options, General tab ... 48
Figure 9-1. Pending Email warning ... 51
Figure 9-2. Unresolved senders. ... 52
Figure 9-3. Resolve Address Dialog ... 53
Figure 9-4. Personal Options (Email) ... 54
Figure 9-5. Allow other uses to send email as you ... 55
Figure 9-6. Email processing for unapproved users and queues ... 56
Figure 9-7. Approve/Reject user email address ... 56
Figure 9-8. Informational message to install the Outlook Client ... 58
Figure 9-9. Disable the display of the Install the CRM Client for Outlook ... 58
Figure 10-1. Search for a company email address ... 61
Figure 12-1. Windows Event Viewer Actions Pane ... 69
Figure 12-2. Windows Event Log Filter ... 70
Figure 14-1. Diagnostics Tool for Dynamics CRM ... 88
Figure 15-1. PFE CRM Trace Tool ... 92
Figure 15-2. Search trace log for errors ... 93
Figure 15-3. Errors found in the trace log ... 94
Figure 18-1. Stop Workflow Step ... 102
Figure 19-1. System Maintenance Job Editor ... 108
Figure 20-1. OrgDbSettings Settings Editor ... 114
Figure 20-2. Edit an Organizational Setting ... 116
Figure 20-3. Organizational Setting Editor Options and Information ... 117
Figure 24-1. Top Queries by Average CPU Time Report ... 131
Figure 26-1. CRM Diagnostics page ... 137
Figure 26-2. CRM Diagnostics page, post run ... 138

Chapter 1.
Create Your Administrative Toolbox

There are several tools that I have found indispensable as a Dynamics CRM Administrator:

Diagnostics Tool for Dynamics CRM

This tool allows you to turn CRM event tracing on or off by making the necessary registry changes.

https://crmdiagtool2011.codeplex.com

CRM Versions: 2011, 2013, 2015, 2016

PFE CRM Trace Tool

Once you have captured your trace logs, this tool allows you to easily view the trace data.

https://pfecrmtracetool.codeplex.com

CRM Versions: 4.0, 2011, 2013, 2015, 2016

Stunnware Tools

This is a toolset from a former Dynamics CRM MVP, Michael Höhne. The primary recommendation here is for the Trace File Viewer functionality.

http://www.donaubauer.com/en/#!StunnwareTools40&slide10

CRM Versions: 4.0, 2011, 2013, 2015, 2016

CRM Maintenance Job Editor

There are several SQL scheduled jobs which are created when Dynamics CRM is installed. This tool allows you to change the default scheduling times. We will discuss this process in the lesson:
Troubleshooting Development Errors.

http://crmjobeditor.codeplex.com

CRM Versions: 2011, 2013, (as of February 1st, 2015, the 2015 version is under developmental review)

Dynamics CRM Organization Settings Editor (OrgDBOrgSettings)

Allows you to edit the internal Dynamics CRM organization settings.

https://orgdborgsettings.codeplex.com

CRM Versions: 2011, 2013, 2015, 2016

Ribbon Workbench

The Ribbon Workbench is a solution that allows for the editing of the Ribbon and Command–bar components found within Dynamics CRM.

http://www.develop1.net/public/page/Ribbon-Workbench-for-Dynamics-CRM-2011.aspx

CRM Versions: 2011, 2013, 2015, 2016

SiteMap Editor

This utility allows you to edit the Dynamics CRM SiteMap, which is the main navigation for Dynamics CRM.

https://www.xrmtoolkit.com/Home/DownloadSitemapEditor

CRM Versions: 2011, 2013, 2015, 2016

XrmToolbox for Dynamics CRM

This is a collection of utilities that perform a variety of functions related to the development and management of Microsoft Dynamics CRM.

http://www.xrmtoolbox.com

CRM Versions: 2011, 2013, 2015, 2016

AstroGrep

This is a text search based on the Unix grep utility and is also very helpful when searching trace files, or any other group of files for that matter, for a specific string or phrase.

http://astrogrep.sourceforge.net

Notepad++

This is a free alternative to using the standard Windows Notepad application and is very good for editing almost any type of text–based files.

http://notepad-plus-plus.org

Dynamics CRM SDK

The Dynamics CRM SDK has a ton of great information for developers as well as administrators. One of the primary tools you may be using is the Plug–in Registration Tool.

You may download the SDK from one of the following links:

Dynamics CRM SDK 2011

http://www.microsoft.com/download/details.aspx?id=24004

The Plug–in Registration tool is found in the bin folder.

Dynamics CRM SDK 2013

http://www.microsoft.com/download/details.aspx?id=40321

The Plug–in Registration tool is found in the Tools\PluginRegistration folder.

Dynamics CRM SDK 2015

http://www.microsoft.com/download/details.aspx?id=44567

The Plug–in Registration tool is found in the Tools\PluginRegistration folder.

Dynamics CRM SDK 2016

https://www.microsoft.com/download/details.aspx?id=50032

The Plug–in Registration tool is found in the Tools\PluginRegistration folder.

NOTES

The Dynamics CRM documentation team produces an update for the SDK about once every six to nine weeks, depending on the actual Dynamics CRM release schedule. I always try and keep at minimum, the prior version, sometimes two or three versions, of each SDK. I do this because there are occasions where an update has introduced an issue with one of the tools included in the SDK and the previous version did not have the issue.

I also number my SDK folders with the version, i.e., SDK-6–0–1, SDK–7.0.0, etc. so that I am clear on the specific version.

This page intentionally left blank.

Chapter 2.
Let's Do a Little Housekeeping

Here are a few things that will make your life as a Dynamics CRM administrator a little bit easier.

Create a CRM Administrators Team

The most efficient method I have found for keeping your entire CRM administration team up–to–date is to create a Team within your organization called **CRM Administrators**.

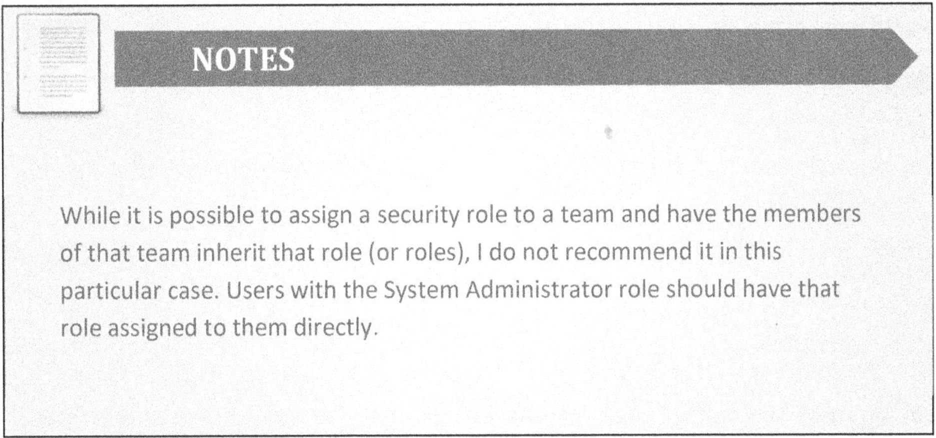

NOTES
While it is possible to assign a security role to a team and have the members of that team inherit that role (or roles), I do not recommend it in this particular case. Users with the System Administrator role should have that role assigned to them directly.

This team will own or has access to all of the management access that we will be creating during this course and it ensures that everyone who needs access, will have access.

Know your system: ISVs, plugins, workflows, custom workflow activates

As a general rule, and for your own sanity, it is a really good idea to document any external applications that interact with your Dynamics CRM system.

This could be custom solutions that have been installed, integration with other systems like Dynamics GP, or custom SSIS packages that move data to and from external systems.

These things are important to know because sometimes a failure in one process or package can lead to errors in another. And any time you have components that are out of your control, they can lead to bigger problems.

Make note of the following:

Custom solutions

Review the list of solutions to see what has been installed from third–party vendors of from internal developers

Plugins and custom workflow activities.

While most of the time plugins and custom workflow activities are deployed using the Dynamics CRM solution model, they do not have to be.

It is important to note which is from a solution and which has been deployed using the plug–in registration tool.

ISV Folder

Review the ISV folder which is located in the Dynamics CRM website folder. This is where custom web pages and code are generally located. This folder is usually empty, but if it is not, then you will need to document where within the Dynamics CRM system any custom pages or other pieces of code appear.

Technically, the usage of the ISV folder was officially discontinued with Dynamics CRM 2013 but many on premises organizations still utilize this folder for custom ASP.NET web page extensions.

Ribbon Modifications

The Ribbon can be modified to add, change, or remove, functionality. It is a really good idea to know and understand what changes have been made. Using the Ribbon Workbench, you can see if customizations have been made which will alter the out–of–the–box Dynamics CRM functionality.

Custom Web Pages

Custom web pages applied as iFrames on individual entity forms or as top–level pages applied to the main navigation.

Everything Else

While you are documenting, it is always a good idea to have the various components modified directly by your organization in your documentation as well. This may not be just code (plugins and JavaScript), but can also include metadata, forms design, workflows, reports, etc.

Tools

Here are some tools that may help with that process:

XRM Toolbox (free)

This toolbox is written and maintained by fellow MVP Tanguy TOUZARD along with several contributors.

http://www.xrmtoolbox.com

SnapShot! for Dynamics CRM (commercial)

Full disclosure: This is a tool that I created and sell. I think it is the best Dynamics CRM documentation tool on the market, but I may be a little biased. Regardless, it creates a ton of documentation about a Dynamics CRM organization. Even the trial version produces more detailed documentation that many other free tools.

http://www.crmaccelerators.net/snapshot

Snagit (commercial)

Snagit is a screen capture tool that I find invaluable for a variety of applications from creating documentation to writing blog articles.

http://www.techsmith.com/snagit.html

Chapter 3.
Error, Warning, and Notification Messages

Dynamics CRM will produce a variety of warning and error messages, which can be categorized into one of the following groups:

Client or Application Errors

Since Dynamics CRM is a web application, there are a variety of errors that can occur at either the Application layer, which is the actual .NET part of the site, or within the JavaScript that is used throughout the application.

Synchronous Platform Errors

The Platform later is what the Dynamics CRM Application actually communicates with to perform the various database operations requested by the user or the system. Synchronous errors occur in the context of a user performing an operation and will usually be shown to the user.

Asynchronous Platform Errors

If you are using workflows, custom workflow activities, or asynchronous plugins, they are all being executed by the Asynchronous Processing Service (instead of the main Dynamics CRM application). Being asynchronous, the user does not normally know these errors occur so you must review the System Jobs to locate and diagnose any errors.

System Notifications

Notifications are either informational or warnings and can be shown to either any user, or to a user with the System Administrator security role, depending on the nature of the notification.

Client or Application Errors

Figure 3-1 shows the dialog that will be displayed when Dynamics encounters an unexpected error within the application layer:

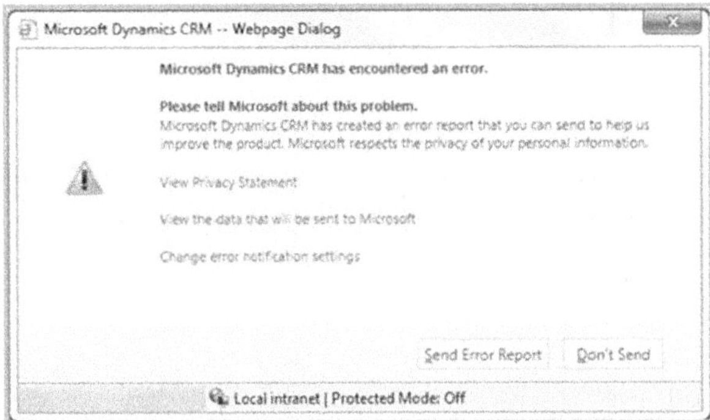

Figure 3-1. Unhandled JavaScript error

You have the option of sending or not sending this information to Microsoft so that they can track the error message in their bug database.

There are also two possibilities as to why/where this error occurred:

- The first is a possible issue with the JavaScript used by Dynamics CRM itself.
- The second is an issue with a piece of custom JavaScript someone has written and installed into your Dynamics CRM installation.

So how do you tell? Just click on the **View the data that will be sent to Microsoft**. This will show you an XML packet containing the offending JavaScript file, line number, and error message captured by the system.

If the file is one of your custom JavaScript web resources, then it is up to you to debug, so click the **Don't Send** button on the dialog. Send that information to one of your developers for further analysis.

If the file does not look like one of your custom web resources, then it is an issue with something inside of Dynamics CRM itself so click **Send Error Report** so that Microsoft can track whatever bug that was encountered.

NOTES

You should train your users to click on the **View the data that will be sent to Microsoft** link, then copy and email you the contents of the box that is displayed. That will help you determine if this is an error within Dynamics CRM, or within some custom code.

Figure 3-2 shows another type of error displayed using JavaScript:

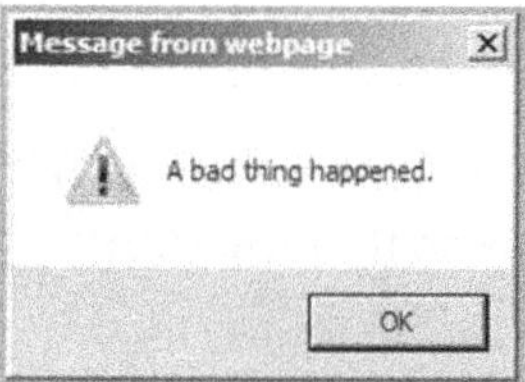

Message from webpage
A bad thing happened.
OK

Figure 3-2. JavaScript error created by a developer

In this particular case, the developer is the one who decided to show this message to the user. (And hopefully this is just a demonstration and not a real error messages – but you know those developers...)

Synchronous Platform Errors

Here are several Dynamics CRM platform errors:

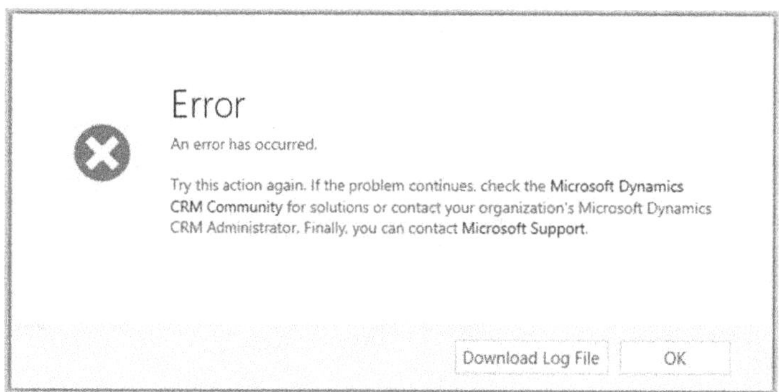

Error
An error has occurred.
Try this action again. If the problem continues, check the Microsoft Dynamics CRM Community for solutions or contact your organization's Microsoft Dynamics CRM Administrator. Finally, you can contact Microsoft Support.
Download Log File
OK

Figure 3-3. Generic Dynamics CRM error (platform layer)

This is a generic Dynamics CRM error that generally is displayed when something happened at the "platform layer," which is where much of the work is done with Dynamics CRM.

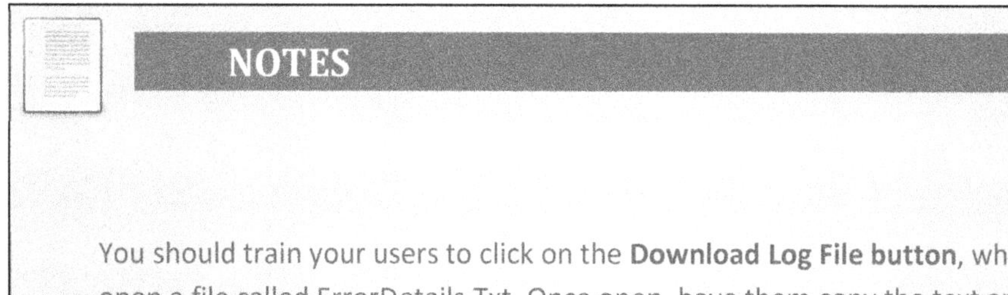

You should train your users to click on the **Download Log File button**, which will open a file called ErrorDetails.Txt. Once open, have them copy the text and email it to you. That will help you determine the origination point for the error. Since this is a platform operation, it could be Dynamics CRM code, or custom code (in a plug–in for instance).

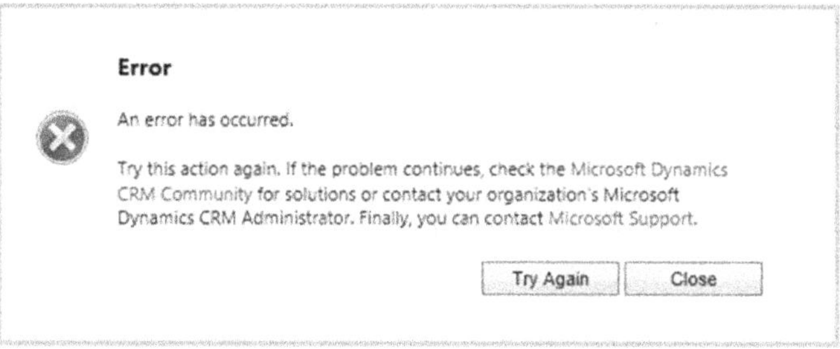

Figure 3-4. Generic Dynamics CRM error (unknown origin)

This error is similar to the one above but without the option to download the log file. This type of error, in many cases, is related to custom web pages, either HTML/JavaScript or ASP.NET. To locate the actual problem that caused this error, you will need to look in the Windows Application event log. We will talk more about that in the lesson:
Monitoring the Windows Event Log.

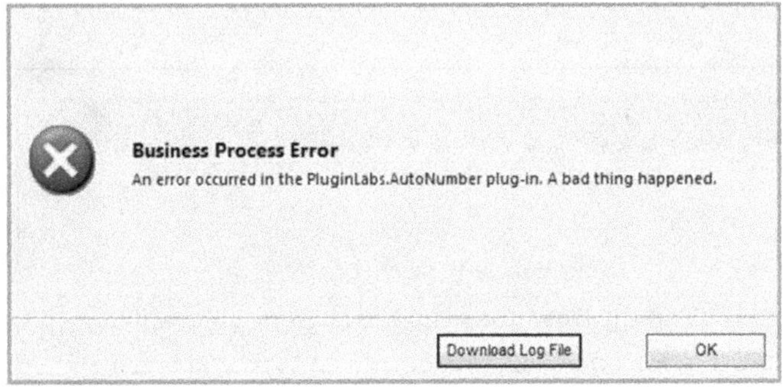

Figure 3-5. Business process error caused by a custom plugin

This is an error displayed from a custom plug–in. You may view additional information on the error by clicking the **Download Log File** button.

Asynchronous Platform Errors

These errors are not presented to the user, in most cases, but rather are added to the system job that was running when the error occurred.

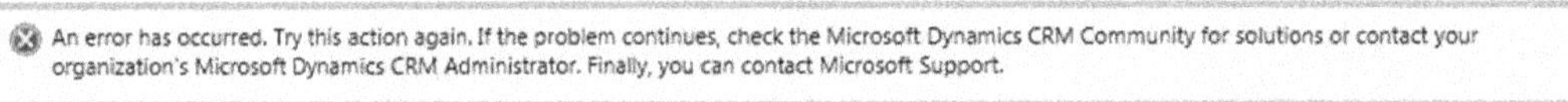

Figure 3-6. Asynchronous plugin error

This is the general error message produced by a plug–in when some type of error has occurred. In many cases, there is not enough information in the error message to determine the actual cause of the error so you must turn on platform tracing to capture the real error message.

We will discuss platform tracing in more detail in the lesson *Platform Event* Tracing Overview.

You cannot send e-mail as the selected user. The selected user has not allowed this or you do not have sufficient privileges to do so. Contact your system administrator for assistance.

Figure 3-7. Error sending email from workflow

This error is usually seen when sending emails via workflow, thought you may also see it if you create an email manually. The error is fairly explanatory and can be attributed to one of two things: Either the user has a security restriction that prevents them from sending email or the user you are attempting to send the email as does not have the personal option, **Allow other users to send email on your behalf** set to yes.

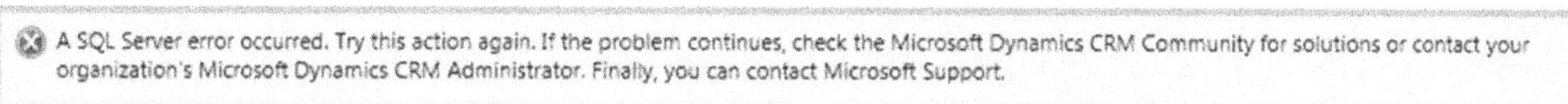

A SQL Server error occurred. Try this action again. If the problem continues, check the Microsoft Dynamics CRM Community for solutions or contact your organization's Microsoft Dynamics CRM Administrator. Finally, you can contact Microsoft Support.

Figure 3-8. Generic SQL Server error

In most cases the SQL Server error is actually a SQL Server operation timeout. Since this is not always the case, you may need to turn on platform tracing to find the real error.

This type of error is also *very* time sensitive. Do not spend a lot of time tracing down this type of error after the fact. In most cases, the error is irrelevant after the fact. If you don't capture the error happening, then you may not be able to locate it later.

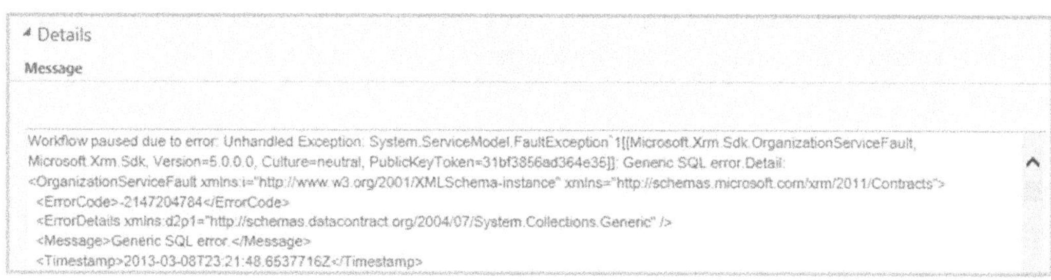

Figure 3-9. System Job error detail

Like the previous error, this type of error is generally a SQL timeout error. The difference between the two error messages is where it occurred. The first error occurred inside of the Dynamics CRM application (or one of its component parts). The second error, the one above, actually occurred inside of a custom workflow activity.

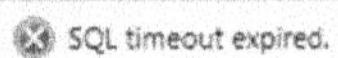

Figure 3-10. SQL Server timeout error

This is a real SQL timeout error which is generally caused by an locking/blocking issue on the database.

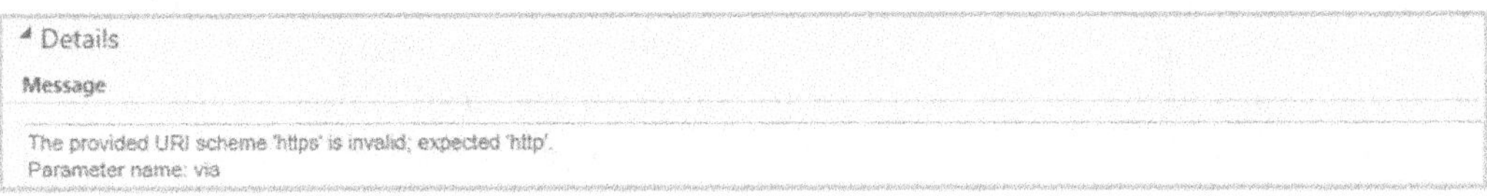

Figure 3-11. Custom plugin error details (configuration)

This error is from a custom plug–in and is the result of an improper piece of configuration information.

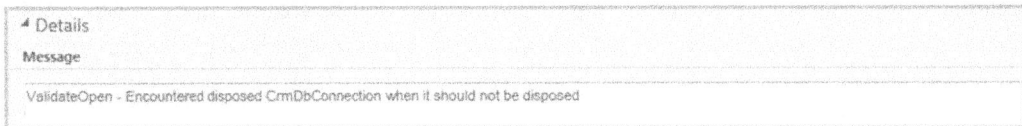

Figure 3-12. Custom plugin error (bad design)

This error occurred inside of a custom plug–in and is a result of a bad design. In layman's terms, the developer requested that a database operation be performed but the connection to the database had already been closed and discarded.

Figure 3-13. Custom plugin error (unhandled exception)

This is an error inside of a custom plug–in.

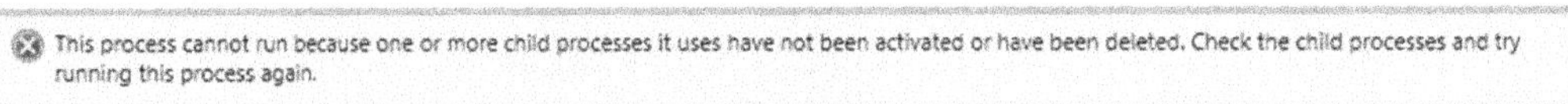

Figure 3-14. Workflow configuration (child workflows) error

This error occurs when you have a workflow that calls additional child workflows. Before the main workflow, or parent, is activated, all of the child workflows that it calls must also be activated.

Figure 3-15. Asynchronous plugin error (non-CRM related timeout)

This error is a little tricky and the issue with trying to diagnose this type of error is, "what operation timed out?" There are few instances within Dynamics CRM were you would see this type of error message, so if you discount that possibility, what is left is an issue with a piece of custom code, which could be a plug–in or a custom workflow activity.

In this particular case, it was a custom workflow activity that added documents to SharePoint. The SharePoint server was having issues and it surfaced inside of Dynamics CRM.

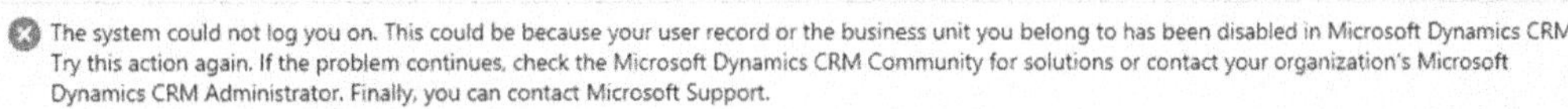

The system could not log you on. This could be because your user record or the business unit you belong to has been disabled in Microsoft Dynamics CRM. Try this action again. If the problem continues, check the Microsoft Dynamics CRM Community for solutions or contact your organization's Microsoft Dynamics CRM Administrator. Finally, you can contact Microsoft Support.

Figure 3-16. User configuration error

This is technically a configuration issue and as the error states, it can usually be one of two things:

- The user has been disabled
- Or, your business unit has been disabled.

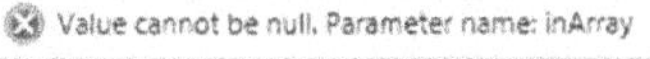

Figure 3-17. Custom plugin error (unhandled exception)

This is an error in a custom plug–in.

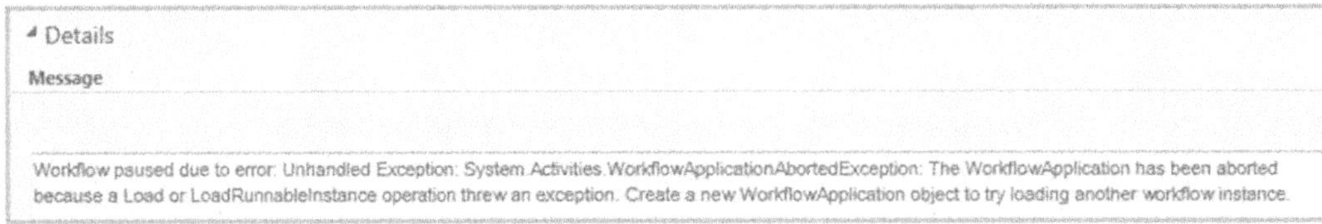
◢ Details

Message

Workflow paused due to error: Unhandled Exception: System.Activities.WorkflowApplicationAbortedException: The WorkflowApplication has been aborted because a Load or LoadRunnableInstance operation threw an exception. Create a new WorkflowApplication object to try loading another workflow instance.

Figure 3-18. Internal Dynamics CRM workflow engine error

This is an error in the Dynamics CRM workflow engine itself. This type of error will either fix itself or may be fixed by restarting the *Microsoft Dynamics CRM Asynchronous Processing Service*.

System Notifications

Dynamics CRM displays a variety of notifications which are generally informational or warning in nature.

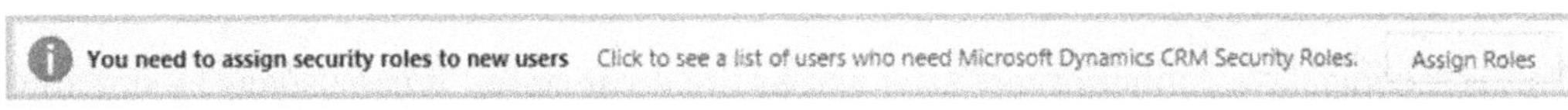

Figure 3-19. Users exist without assigned security roles

This is an informational message (as you can see by the icon). Users with the System Administrator role will see this message when they first log into Dynamics CRM. To assign security roles to these users, just click the **Assign Roles** button and you will be taken to a special User view where you can see a list of Dynamics CRM users that do not have security roles assigned.

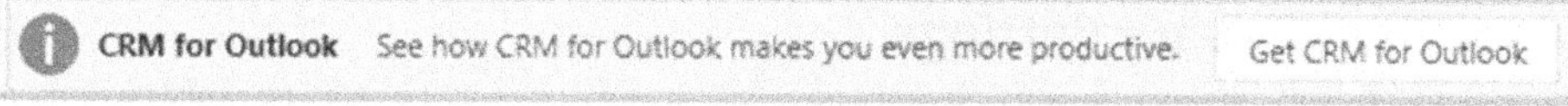
CRM for Outlook
See how CRM for Outlook makes you even more productive.
Get CRM for Outlook

Figure 3-20. Informational message to install the Outlook Client

One of the features of Dynamics CRM is to inform the user if they have Microsoft Outlook installed but do not have the Dynamics CRM Client for Outlook installed. The message is displayed at the top of the Dynamics CRM window as shown in Figure 3-20.

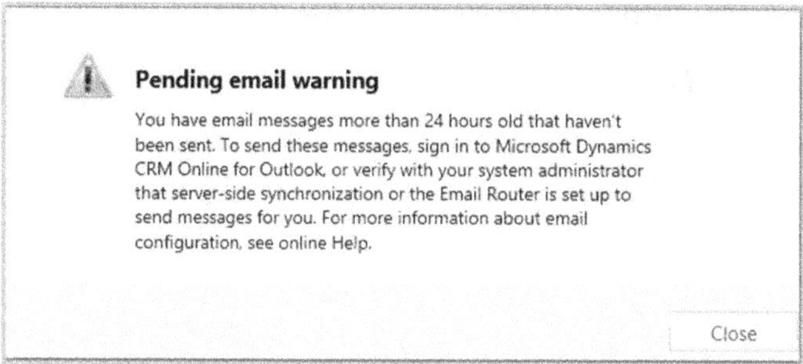
Pending email warning
You have email messages more than 24 hours old that haven't been sent. To send these messages, sign in to Microsoft Dynamics CRM Online for Outlook, or verify with your system administrator that server-side synchronization or the Email Router is set up to send messages for you. For more information about email configuration, see online Help.
Close

Figure 3-21. Pending Email warning

Emails which are in a Pending state are those that are waiting to be sent using either the Email Router, the Microsoft Dynamics CRM Client for Outlook, or using Server Side Synchronization.

If pending emails exist, any user with the System Administrator security role will see the alert shown in Figure 3-21, when they first open the Dynamics CRM web site.

Error Report and Privacy Settings

Believe it or not, reporting of errors to Microsoft actually falls under the Privacy Settings. This allows you to set, at the system or individual level, whether or not the "send error to Microsoft" is displayed to the user.

There are two types of privacy settings:

- System–wide, found within the Settings, Administration area
- Personal, found on each user's Personal Options, Privacy page

Figure 3-22 shows the system–wide settings. If this setting is not enabled, each user will have the ability to make their own decision about how error reporting should be handled.

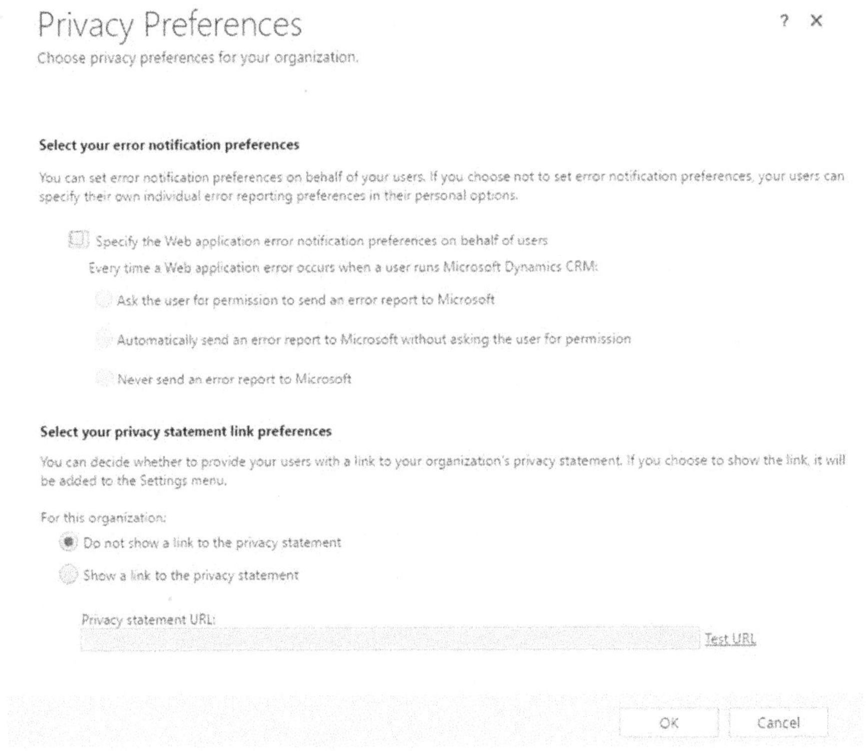

Privacy Preferences

Choose privacy preferences for your organization.

Select your error notification preferences

You can set error notification preferences on behalf of your users. If you choose not to set error notification preferences, your users can specify their own individual error reporting preferences in their personal options.

Specify the Web application error notification preferences on behalf of users

Every time a Web application error occurs when a user runs Microsoft Dynamics CRM:

Ask the user for permission to send an error report to Microsoft

Automatically send an error report to Microsoft without asking the user for permission

Never send an error report to Microsoft

Select your privacy statement link preferences

You can decide whether to provide your users with a link to your organization's privacy statement. If you choose to show the link, it will be added to the Settings menu.

For this organization:

Do not show a link to the privacy statement

Show a link to the privacy statement

Privacy statement URL:

Test URL

OK Cancel

Figure 3-22. System-wide privacy settings

Figure 3-23 shows the options available to the user regarding error reporting.

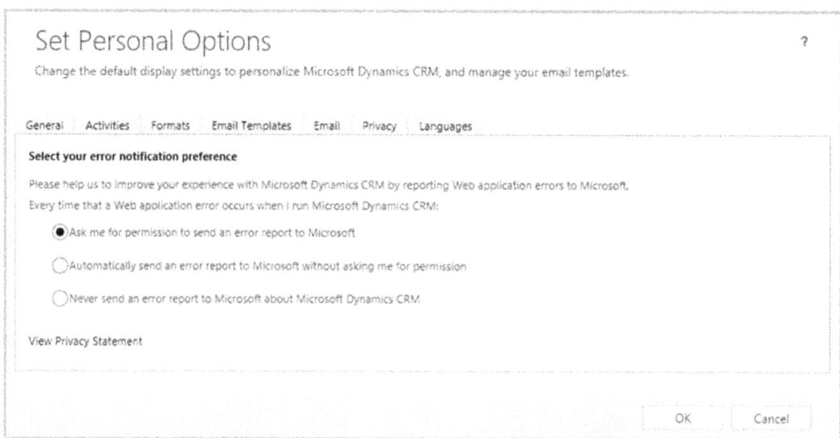

Figure 3-23. Personal privacy settings

this page intentionally left blank

Chapter 4.
Managing System Jobs

The management of System Jobs is one of the most important functions of any Dynamics CRM administrator. Issues that appear in the system jobs list can sometimes be indicative of a systemic problem affecting all users; so it is always a good idea to monitor the system jobs on a frequent basis.

There are two kinds of System Jobs within Dynamics CRM:

System Initiated

These jobs are started by the system to handle various internal operations such as maintenance and cleanup.

User Initiated

These jobs are the result of some user action starting a process such as a workflow or firing an asynchronous plug–in.

Depending on the size of your user–base and the daily activity, you can add hundreds, thousands, or tens of thousands of rows to system job table every day.

So why is this so bad?

Well, it depends on your hardware and usage, but I have seen 50,000 records reduce the throughput of a Dynamics CRM system to the point where we had to turn off the Asynchronous service just so users could use the system.

At the opposite end of the scale, I have seen other systems with millions of rows that did not seem to affect anyone. Again, it really depends a lot on your hardware and how you are using Dynamics CRM.

We will discuss system job cleanup in *Automating System Job* Cleanup.

Problem Indicators

There are a range of problems related to system jobs that can occur within your Dynamics CRM system. Here are some things to look for:

Waiting System Jobs

While there are instances where a system job can be in a waiting state, there are times when a waiting job is actually a failed job. If the Message field contains any data, then it is probably a failed job. Sometimes waiting jobs that have failed can be restarted using the Resume action, but this will totally depend on the type of failure.

Failed System Jobs

Failed jobs generally result from an unrecoverable system error – but not always. Failed jobs can also be resumed, if the failure is not permanent. It is hard to judge if the failure is permanent until you try the resume action and it either works or it gives you an error.

The Asynchronous Service is not running

There should be *some* type of system job processed every few minutes. If you navigate to system jobs and notice that the last **Started On** time stamp was more than an hour or two old, that is an indication that the Asynchronous Service is not running. I talk more about this in *Regularly check to ensure that workflows and system jobs are running*.

System Job Actions

The System Jobs view is not just for looking at your current and past system jobs. As you can see in Figure 4-1, you may also alter the state of a system job, depending on the current state.

Figure 4-1. System job actions

These actions are either displayed on the toolbar, or below the Actions button, depending on whether you are looking at the system jobs view, or have an individual system job record open.

Actions may be performed on a single or multiple system jobs. Should an error occur when performing an action against multiple system jobs, a generic error message as shown in Figure 4-2, will be displayed.

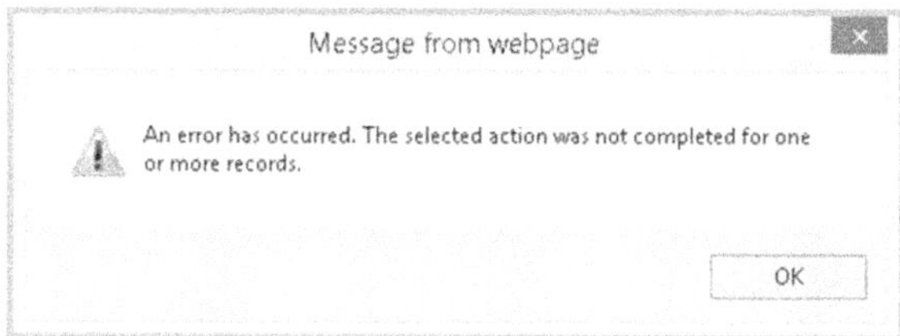

Figure 4-2. System job action failure - multiple records

In order to locate the system job with an error, you will need to reduce the number of records you are performing the action against until you find the system job causing the failure. Keep in mind that there may be more than one record preventing a successful action.

Cancel

Cancels a system job.

Pause

Pauses a system job that is in a running (or waiting) state.

Resume

Resumes a system job. After selecting the record or records, a confirmation dialog will be displayed, as shown in Figure 4-3.

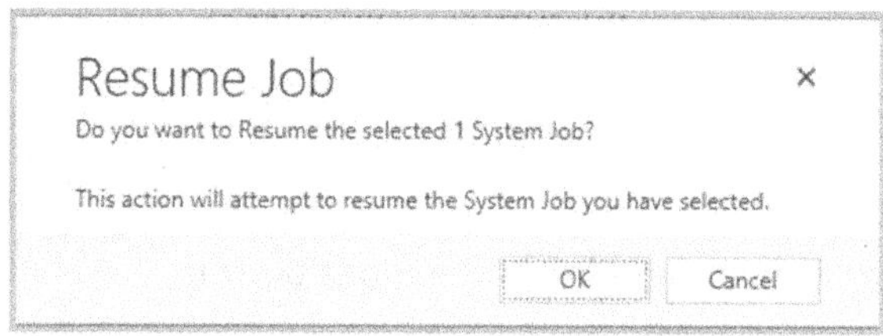

Figure 4-3. Resume System Job

Notice that it shows the number of system jobs that the action will be performed on. Upon clicking OK, the resume action will be applied to all of the selected records.

Should any records fail to resume, the dialog in Figure 4-4 will be shown.

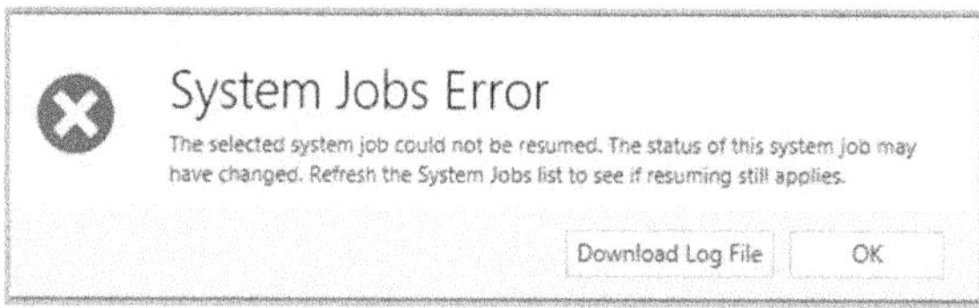

Figure 4-4. Failure to resume a system job

Note: You cannot resume a canceled system job.

The ability to resume a system job will depend on why it stopped in the first place. If it was data or configuration–related, then then the probability is high. If the system job stopped because of environmental issues, such as a point–in–time system slowdown or failure, then those too may be resumed.

The biggest question is, "Do I need to resume this system job?"

If the system job was performing a function vital to the outcome of the specific record or internal task, then the answer may be yes, but in other instances you may not wish the job to be restarted. Again, it is a situation issue that you will need to use your judgment and experience to answer.

Postpone

Postpones the current system job. You can specify a date and time, as you can see in Figure 4-5.

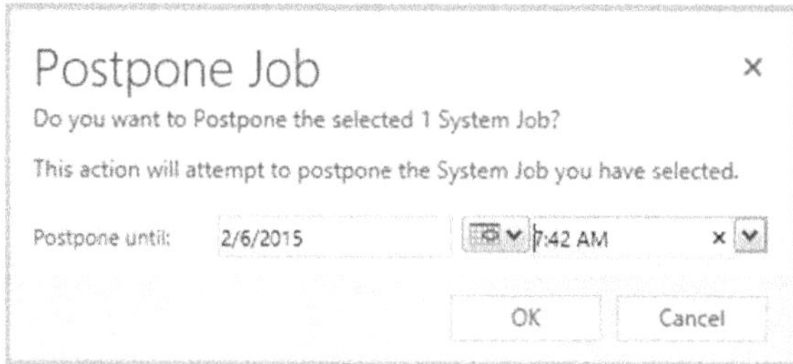

Figure 4-5. Postpone a system job

Workflow Job Retention

As shown in Figure 4-6, the author of a Dynamics CRM workflow has the option of deleting the system job if the workflow ran successfully.

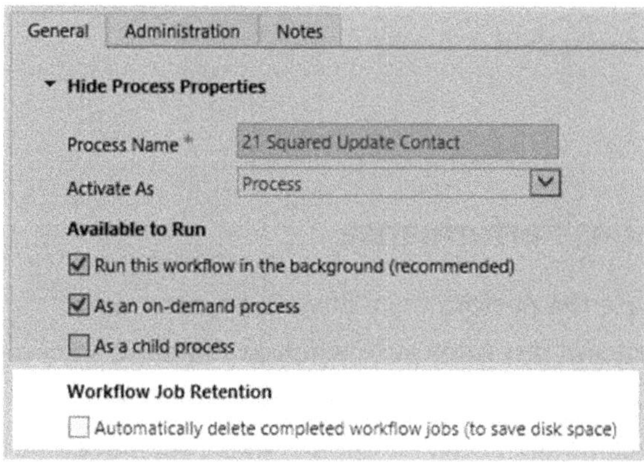

Figure 4-6. Workflow job retention

On general principle, I advise this setting only be used in cases where the CRM administrator does not require evidence that a particular workflow actually ran. If you delete the system job associated with a successful workflow, then you never have proof that the workflow ran so troubleshooting after–the–fact is very difficult.

I prefer to perform a bulk–cleanup of successful workflows at a periodic interval. We will be discussing this topic in–depth in
Automating System Job Cleanup.

Plug–ins: Delete AsyncOperation if StatusCode = Successful

Asynchronous plug–ins have a similar setting that will allow the developer to automatically remove the system job associated with the running of the plug–in step if it was successful, as shown in Figure 4-7.

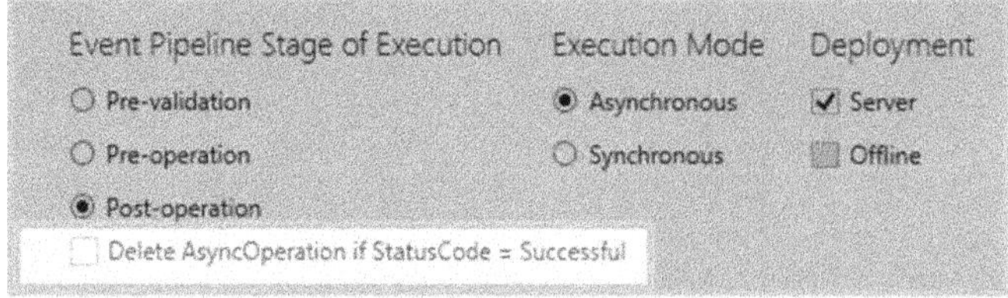

Figure 4-7. Delete AsyncOperation if StatusCode = Successful

But like the workflow system jobs, there are times when you need this evidence. If this is the case, then do not check this box. Otherwise, this is a great way to keep your system job table from containing irrelevant records.

We will be discussing the cleanup topic in–depth in
Automating System Job Cleanup.

Increase Asynchronous Operations Performance

There are a several things that can increase the performance of your Asynchronous Service. There are, unfortunately, no hard and fast rules as to when you need to perform such actions because everything will depend on how your Dynamics CRM system is utilized, how your hardware is configured, and how many users you have.

Move the Asynchronous Service

By default, the asynchronous service is installed on the Dynamics CRM web server. This can lead to contention issues as both try to utilize the same resources. If you move the asynchronous service to a server of its own, you may find that both the web server and the asynchronous server have improved performance.

Install Multiple Asynchronous Servers

You can install multiple instances of the Asynchronous Service, but you should probably not add more than four (4). In most cases, two will handle the processing without issue, but again, this will totally depend on your situation.

As with the previous note, the key is to *not* have an asynchronous service running on the Dynamics CRM web server.

this page left intentionally blank

Chapter 5.
System Job Management: Registry and Database Settings

Here are a few tips and tricks for helping increase the performance of your Dynamics CRM Asynchronous Service

How to improve Microsoft CRM Async Performance on multi-core servers

Note: This article was written for Dynamics CRM 4.0 and this setting should already exist, should you have version 2011 or higher, but it would not hurt to verify.

Common queries regarding Dynamics CRM Asynchronous Service

This is a really great article by the EMEA Dynamics CRM support team.

AsyncRemoveCompletedJobs

AsyncRemoveCompletedWorkflows

These two Microsoft KnowledgeBase articles describe registry keys that can be used to enhance performance of your Dynamics CRM system by automatically deleting completed workflows and system jobs and.

this page left intentionally blank

Chapter 6.
Creating System Management Views

I like to start by creating a number of personal views which will help filter out the irrelevant data.

There are several reasons to use personal views over system views for many management functions within Dynamics CRM:

1. Not all entities, like System Jobs, support custom system views.
2. Personal views do not fall within the Dynamics CRM solution system and therefore are not packaged with other customizations. This allows the views to be maintained by the system administrators rather than the developers, so changes made to these views are not subject to your company's normal development cycle.

System Jobs with Messages

The message field contains any error messages that occur during the execution of the system job. This field is always blank unless an error occurred. So, to search for System Jobs that have encountered errors, we need to create a personal view whose criteria matches Figure 6-1.

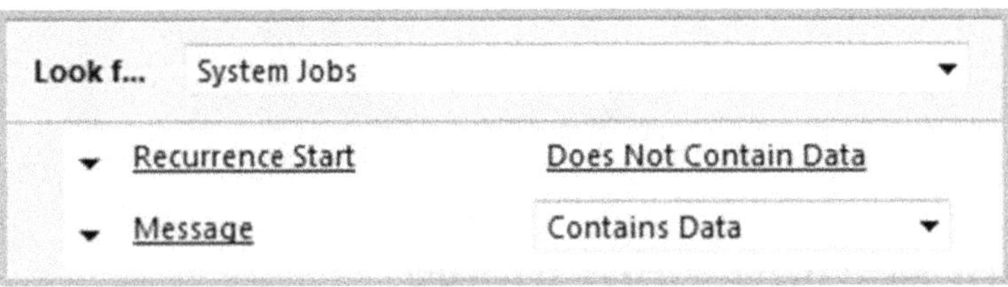

Figure 6-1. System Jobs with Messages

System Jobs with Messages (Recent)

The next personal view is a modification of the System Jobs with Messages view but with a date restriction added, as you can see in Figure 6-2.

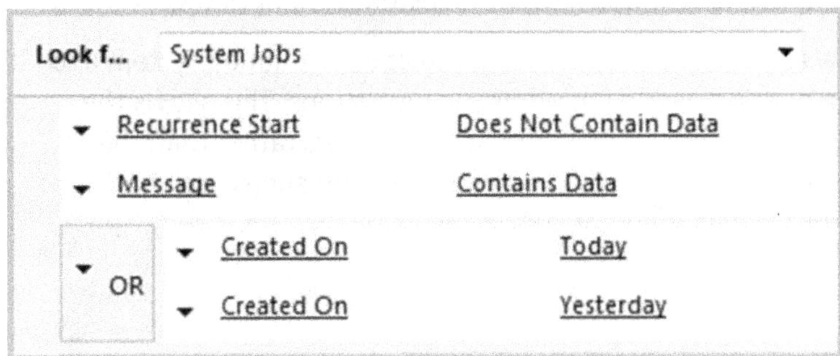

Figure 6-2. System Jobs with Messages (Recent)

As we discussed earlier, most of the time a review of the System Jobs data is only relevant when you are having issues. By limiting your search to just the current and previous day, you (hopefully) prevent yourself from getting lost in the weeds and trying to track down errors or conditions that no longer exist.

System Jobs (Canceled or Failed)

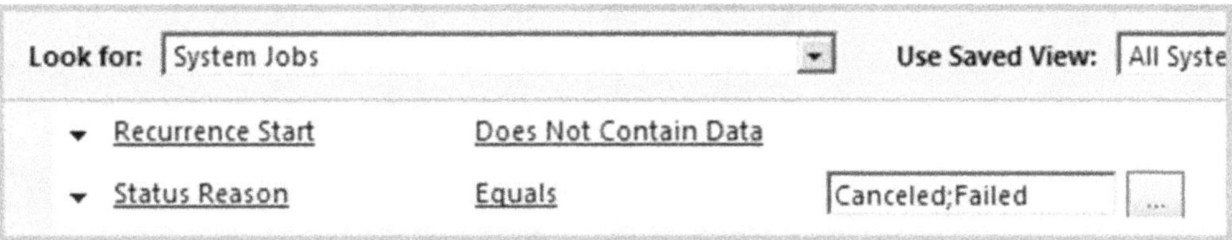

Figure 6-3. System Jobs (Canceled or Failed)

This view is fairly important because it shows not only system jobs that have failed outright, but also jobs that may have been cancelled by either the system or by a user. The failed jobs should have a reason for their failure listed in the details section of the system job (in the message field) while the Cancelled jobs should have the name of the person who canceled the job listed as the last modified by user.

this page left intentionally blank

Chapter 7.
Automating System Job Cleanup

There are two methods to automate the system job cleanup:

- SQL script
- CRM bulk delete job

These options are not mutually exclusive, so let's discuss when and where each option is useful.

NOTES

This type of operation is one of the few instances that Microsoft actually recommends or supplies information regarding working directly with the SQL database. The information presented here may be validated using the following KnowledgeBase article: http://support.microsoft.com/kb/968520

Cleanup Using SQL Script

I generally use SQL Scripts to clean up system jobs when I first start working on a Dynamics CRM installation that is new to me. Chances are, the system jobs have never been maintained and there could be hundreds of thousands or even millions of rows of data to delete.

Such a massive cleanup can be better handled by a SQL script than the second method, a bulk delete job.

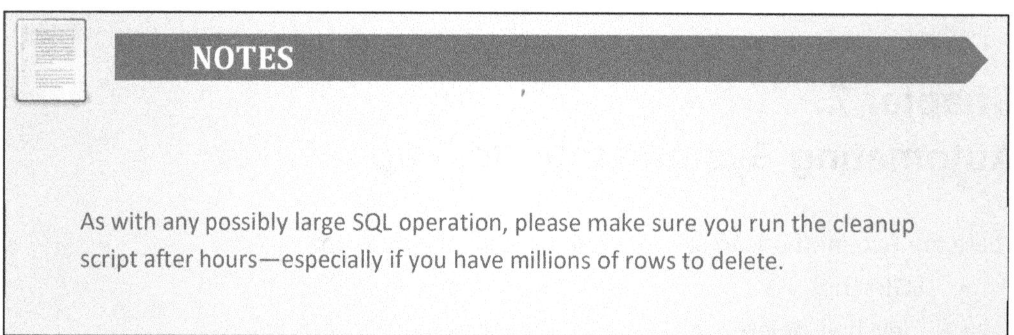

As with any possibly large SQL operation, please make sure you run the cleanup script after hours—especially if you have millions of rows to delete.

There are actually three SQL scripts involved in this process:

- A script to count the number of rows that will be deleted
- A script to create indexes that will speed up the delete operation
- The main script that actually deletes the records that match a specific criterion

The create index script generally only has to be run once per installation and the count script can be run any time you need to see how much data will be cleaned up.

The main cleanup script can be run periodically either manually, or as part of a SQL job that gets executed daily, weekly, or monthly, depending on your requirements.

Filter Criteria Explained

The main cleanup script using three filters to determine what records to delete. Default values for each field are **highlighted**:

OperationType

- **1 System Event**
- 2 Bulk Email
- 3 Import File Parse
- 4 Transform Parse Data
- 5 Import
- 6 Activity Propagation
- 7 Duplicate Detection Rule Publish
- 8 Bulk Duplicate Detection
- **9 SQM Data Collection**
- **10 Workflow**
- 11 Quick Campaign
- **12 Matchcode Update**
- 13 Bulk Delete
- 14 Deletion Service
- 15 Index Management
- 16 Collect Organization Statistics
- 17 Import Subprocess
- 18 Calculate Organization Storage Size
- 19 Collect Organization Database Statistics
- 20 Collection Organization Size Statistics
- 21 Database Tuning
- 22 Calculate Organization Maximum Storage Size
- 23 Bulk Delete Subprocess
- 24 Update Statistic Intervals
- **25 Organization Full Text Catalog Index**
- 26 Database log backup
- **27 Update Contract States**
- 28 DBCC SHRINKDATABASE maintenance job
- 29 DBCC SHRINKFILE maintenance job
- 30 Reindex all indices maintenance job
- 31 Storage Limit Notification
- 32 Cleanup inactive workflow assemblies
- 35 Recurring Series Expansion
- 38 Import Sample Data
- 40 Goal Roll Up
- 41 Audit Partition Creation
- 42 Check For Language Pack Updates
- 43 Provision Language Pack
- 44 Update Organization Database
- 45 Update Solution
- 46 Regenerate Entity Row Count Snapshot Data
- 47 Regenerate Read Share Snapshot Data
- 50 Outgoing Activity
- 51 Incoming Email Processing
- 52 Mailbox Test Access
- 53 Encryption Health Check
- 54 Execute Async Request
- 49 Post to Yammer
- 56 Update Entitlement States

StateCode

0	Ready
1	Suspended
2	Locked
3	**Completed**

StatusCode

0	Waiting For Resources
10	Waiting
20	In Progress
21	Pausing
22	Canceling
30	**Succeeded**
31	Failed
32	**Canceled**

Cleanup Using Bulk Delete

As a standard maintenance process, I create a series of Bulk Delete jobs that will clean up the system jobs table on a frequent basis. The frequency depends on the amount of activity being generated so you will need to take my advice and apply it to your own organization.

Cleanup Job #1

The first bulk delete job you need is one to remove any workflow jobs that were successfully completed and which are over one month old:

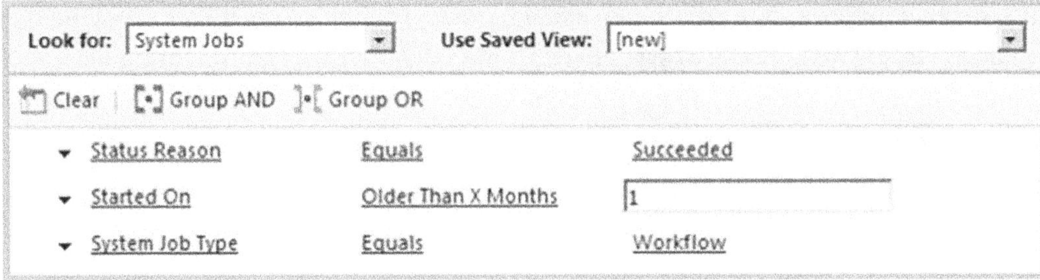

Figure 7-1. Bulk delete successful workflows

This bulk delete is very similar to the functionality of the SQL script we used to perform the initial cleanup.

I generally set this job to run once per week, starting on a weekend night so as not to interfere with production. As mentioned, you may need to alter any of the bulk delete job's parameters or the run frequency, depending on your environment.

Cleanup Job #2

The next bulk delete job removes any system jobs, over one month old, with a status of Canceled of Failed:

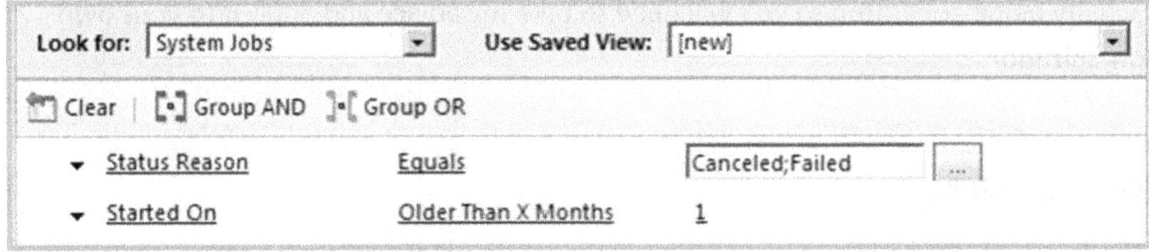

Figure 7-2. Bulk delete canceled and failed system jobs

If a system job is over one month old, and it has had a hard–failure, it is of very little value to me so I remove any of system jobs that match that criteria. This also helps reduce the clutter when reviewing the system jobs using the personal view, **System Jobs (Canceled or Failed)**.

Cleanup Job #3

This bulk delete job will remove any system jobs that contain error messages that are over one month old:

Figure 7-3. Bulk delete system jobs with messages

Chapter 8.
User Management Tips and Tricks

Before we begin this lesson I want to be clear that we will not be covering the actual management of Dynamics CRM users, but rather we will focus on tips and tricks that you may find useful.

Built–In System Views

The following system views provide useful functionality to the CRM Administrator.

Disabled users consuming licenses

I am not totally sure if this is a valid scenario, but the view does exist. Theoretically, when you disable a user, the license consumed by that user is returned to the pool of licenses. In the case that this did not happen, you can use this view to see if any such users exist.

To correct such a situation, you would perform the following steps:

1. Open the disabled user.
2. Re–enable the user.
3. Change their license type to Administrative, as you can see in Figure 8-1.

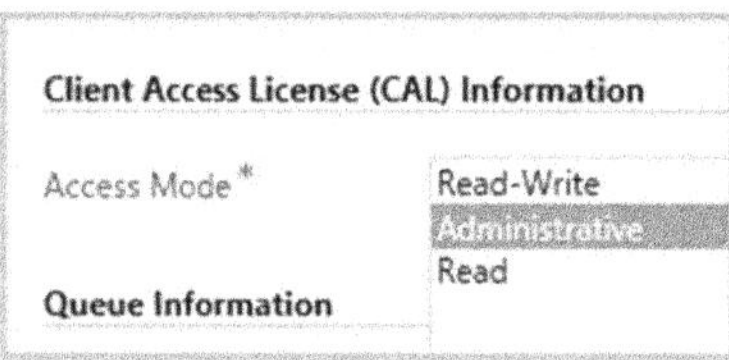

Figure 8-1. Client Access License Information

4. Save the user record.
5. Disable the user.

Users with no assigned security roles

This view will contain a list of all Dynamics CRM users who do not have at least one security role defined. This is also the view that you are displayed when you click on the Assign Roles button when you see the notification shown in Figure 8-2:

Figure 8-2. Users exist without assigned security roles

Users: Primary E-Mail (Pending Approval)

This view shows us users whose primary email awaits approval. Depending on how your system was configured, you may need to manually approve a user's email address before it can be used to send and receive emails.

Note: We will actually be covering how to use this view in–depth in Chapter 9.

Enabled Users

I have found it useful to add both the **Manager** and **Position** fields to this view. This is extremely helpful should you ever implement hierarchical security.

Personal Views

Creating the following personal views can give you a bit of a better view into the management of your users.

Enabled Users (Detail)

One of the first views I create for users is a modification of the Enabled Users view, but with extra detail, which includes email and licensing information. This is very useful when you need a quick glance at the overall user community.

Make your column layout resemble this:

- Full Name
- Primary E-mail
- E-mail 2
- Business Unit
- Access Mode
- License Type
- Restricted Access
- Incoming E-mail Delivery Method
- Outgoing E-mail Delivery Method
- Primary E-mail Status

Personal Options

While there are many personal options available, figure Figure 3-1 shows the settings that can are most commonly modified, or which can cause you the most pain and suffering if set incorrectly.

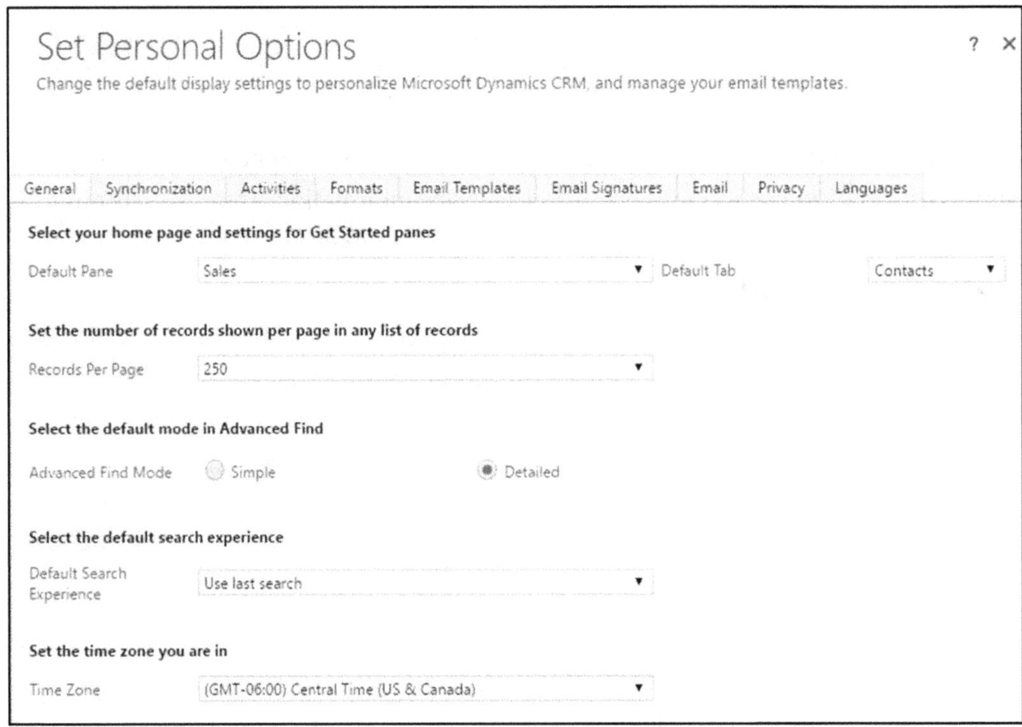

Figure 8-3. Personal Options, General tab

Select your home page and settings for Getting Started panes

Allows the user to set how Dynamics CRM will appear when first opened.

Records per page

The default is 50, which is usually the setting you wish users to have. If you get complaints about speed, you may find that someone has altered the setting and is now showing more records per page, when viewing data. This has a tendency to slow down the response time, strictly because it is retrieving more data.

Time Zone

Time zones within Dynamics CRM can be quite problematic, if set incorrectly. Unless you manually change it, dates are always stored in Universal Coordinated Time (UTC) in the database.

When a user, or an application, requests data from the database, it is automatically converted into the user's local time zone.

If the time zone is incorrect, this can lead to confusion on the user's part and confusion leads to phone calls. And we don't like phone calls, do we?

So, make sure the user's time zone is set to their physical location.

this page intentionally left blank.

Chapter 9.
Email Management

Pending Email Warning

Emails which are in a Pending state are those that are waiting to be sent using either the Email Router, the Microsoft Dynamics CRM Client for Outlook, or using Server Side Synchronization.

If pending emails exist, any user with the System Administrator security role will see the alert shown in Figure 9-1, when they first open the Dynamics CRM web site.

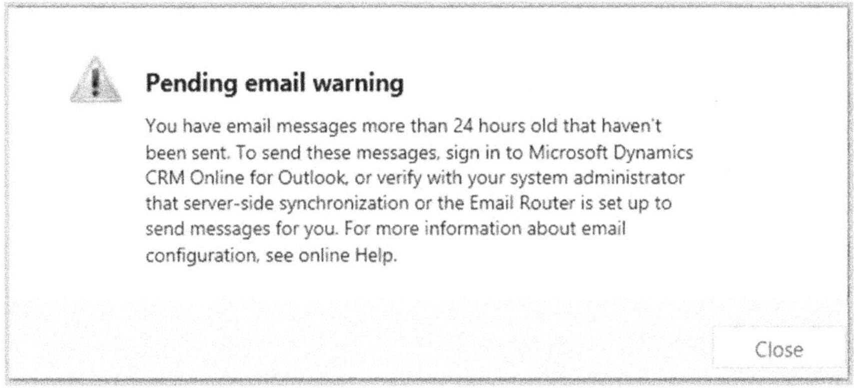

Pending email warning
You have email messages more than 24 hours old that haven't been sent. To send these messages, sign in to Microsoft Dynamics CRM Online for Outlook, or verify with your system administrator that server-side synchronization or the Email Router is set up to send messages for you. For more information about email configuration, see online Help.
Close

Figure 9-1. Pending Email warning

In some cases, there are legitimate reasons for email to be in a Pending state, but most of the time it is an indication of either a misconfiguration or a service is not running. Here are the usual suspects:

Outlook

A user has their Outgoing Email set to Microsoft Outlook and:

- Outlook is not running. (maybe they are on vacation)
- Or the user was prompted to send the email through Outlook and they clicked **No**
- The Dynamics CRM Client for Outlook is not installed

Email Router

The Microsoft CRM Email Router service is not running or it is over–taxed and taking an extremely long time to send email.

Server–Side Sync

If you are using Server–Side Sync to synchronize your mailboxes there could be instances where the mail is not flowing. Since Server–Side uses the Microsoft Dynamics CRM Asynchronous Service to actually process the mail, then that service may not be running or needs to be restarted.

E-mails with Unresolved Senders

What, you may ask, is an unresolved sender? Well, it looks like Figure 9-2:

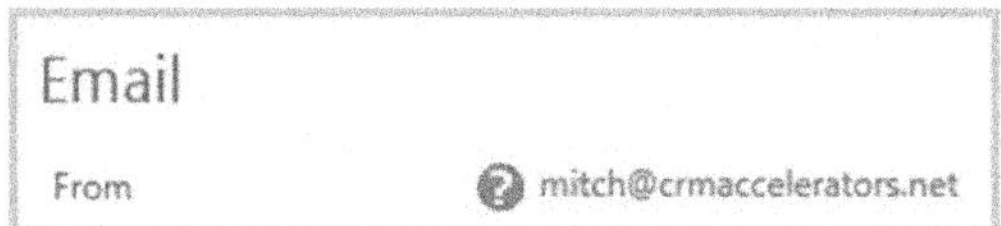

Figure 9-2. Unresolved senders.

This is caused when an email is added to Dynamics CRM and there is not a record associated with the email. Is this a big deal? Well, that would really depend on your organization and how you like to handle email. But in most cases, the answer is no.

By the way, to correct an unresolved email, just click on the email, and the dialog shown in Figure 9-3 will be displayed:

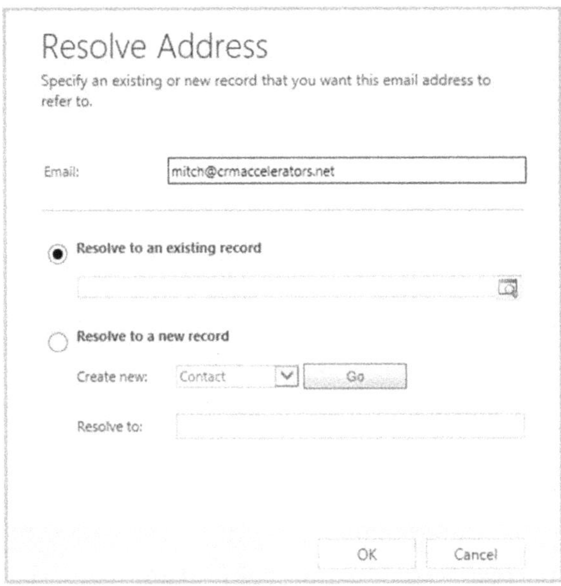

Figure 9-3. Resolve Address Dialog

As you can see, you have the ability to resolve to an existing record, or allows you to create a new record. Unfortunately, this is a manual operation that must be performed on each email.

User Personal Options (Email)

There is one setting within a user's personal options that can be considered quite dangerous. I consider it dangerous because:

1. Few people know the setting exists.
2. It is turned on by default.

The setting is called, **Automatically Create Records in Microsoft Dynamics CRM**, as you can see in Figure 9-4:

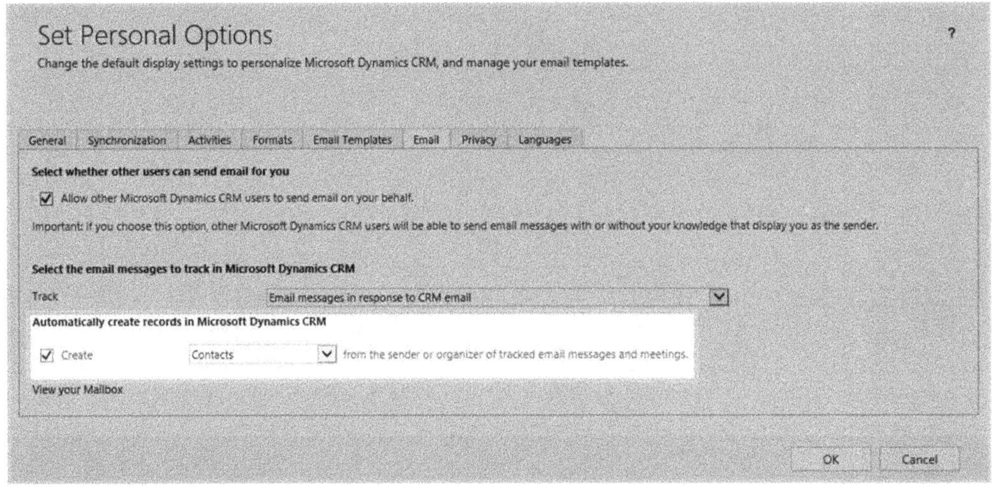

Figure 9-4. Personal Options (Email)

The default setting is to automatically create a new contact should an email be added to Dynamics CRM with an email address that could not be resolved to an existing record.

If this is an undesired activity, then you will need to disable this feature for each user.

Another setting that can sometimes be a problem is the setting shown in Figure 9-5. Allow other uses to send email as you.

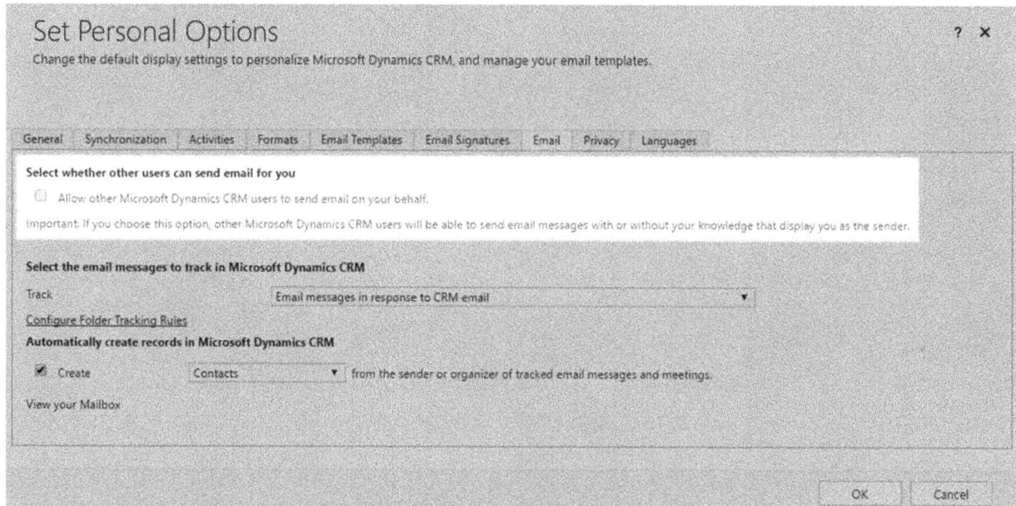

Figure 9-5. Allow other uses to send email as you

If this setting is not enabled, then any automation you may have (workflows) will not be able to send emails as the user.

Approved Email for Users and Queues

Starting in Dynamics CRM 2011, a feature was added that allowed the system administrator to specify that emails for users and queues must be manually approved before being processed by the email router.

This sometimes caused a problem because the default setting was set to *on*, depending on the version of Dynamics CRM that was in use. This caused all sorts of issues debugging email transactions.

There are two parts to this feature. The first is the system setting, as you can see in Figure 9-9.

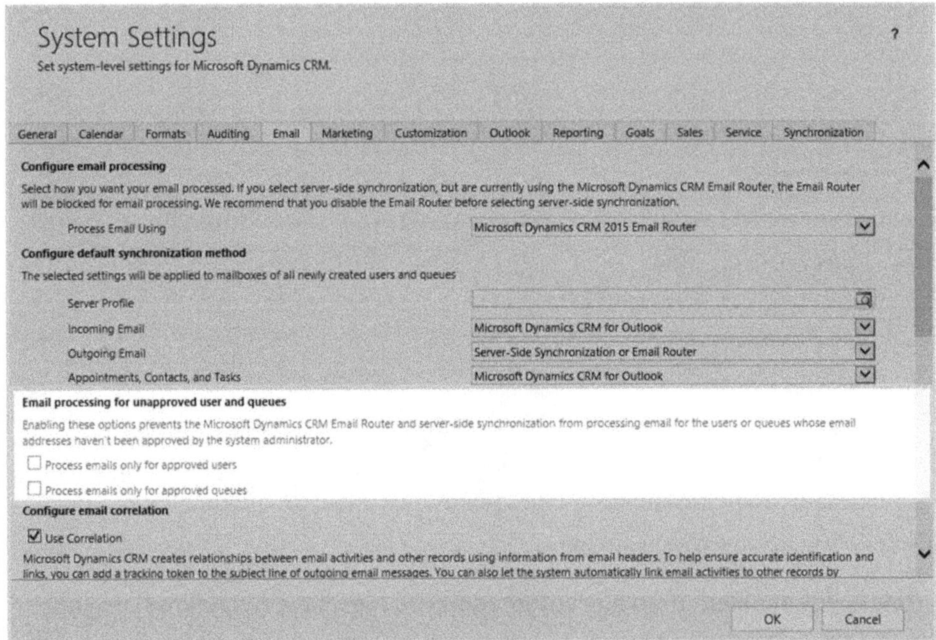

Figure 9-6. Email processing for unapproved users and queues

You can activate approval for either users, queues, or both.

The secondary process is to manually approve one of these entities. If we open a user, you will see the Approve and Reject buttons on the Command Bar shown in Figure 9-7.

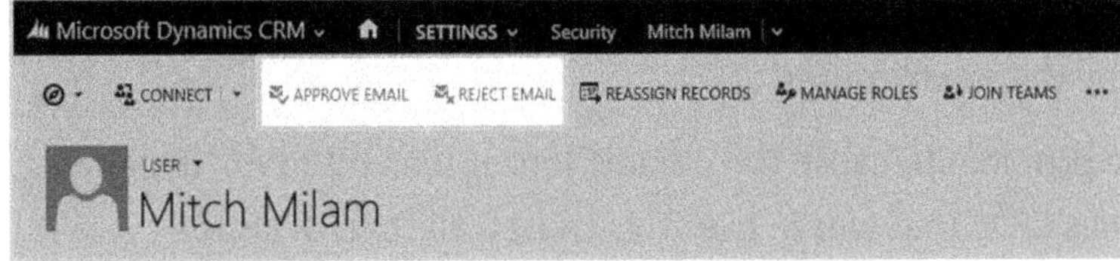

Figure 9-7. Approve/Reject user email address

The Queue entity has the same buttons on the Command Bar.

Why is this a Problem?

The biggest issue with this setting is that you have no idea when a user has been approved or not. This information is actually stored in the field, **Primary Email Status**, which is not on the user or queue forms by default so the setting is not obvious.

What is worse, is if this setting is turned on and a user or queue has not been approved, the email router will totally ignore their email address for any processing.

This can lead to a lot of wasted time until you track it down.

Install the Dynamics CRM Client for Outlook

One of the features of Dynamics CRM is to inform the user if they have Microsoft Outlook installed but do not have the Dynamics CRM Client for Outlook installed. The message is displayed at the top of the Dynamics CRM window as shown in Figure 9-8:

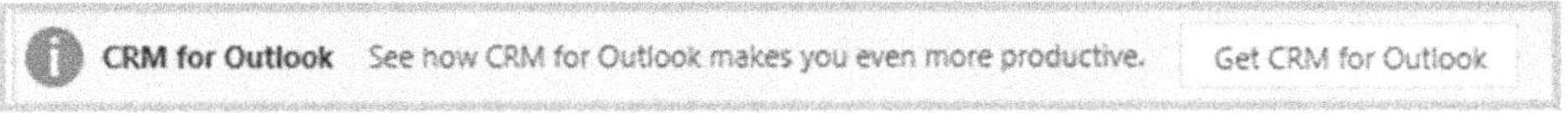

Figure 9-8. Informational message to install the Outlook Client

In many cases, you do not wish the user to install new software on their workstation. Luckily, there is a System Setting that allows us to disable that message, as you can see in Figure 9-9.

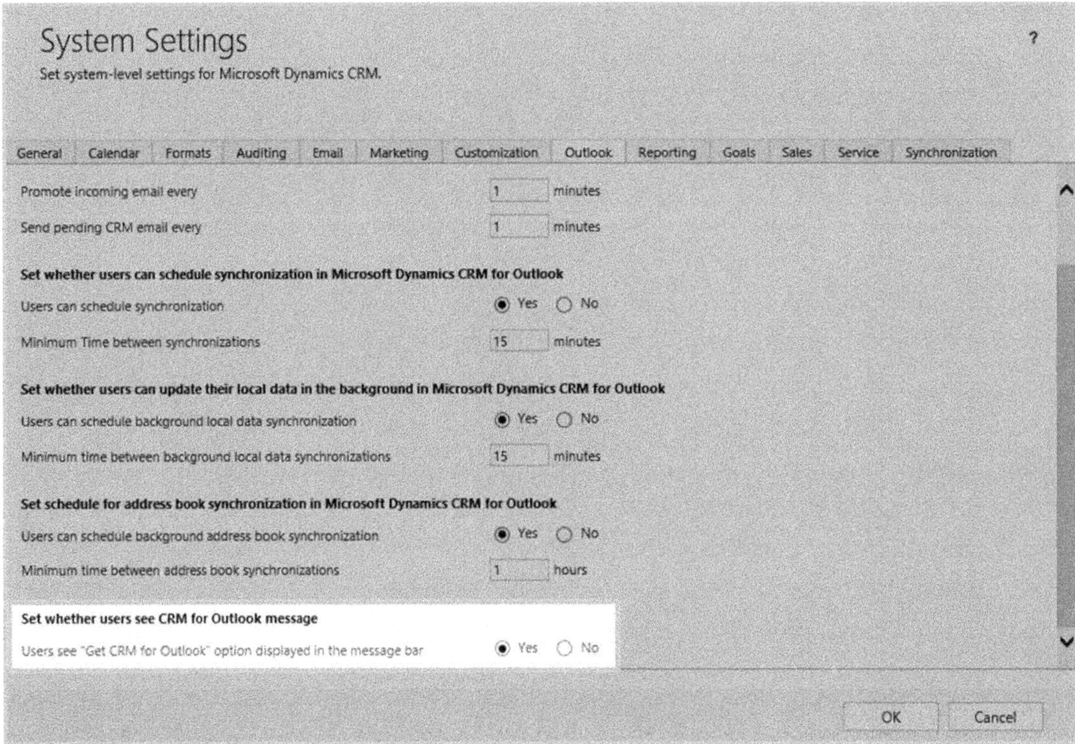

Figure 9-9. Disable the display of the Install the CRM Client for Outlook

Chapter 10.
Email Management – Preventative Maintenance

There is one aspect of Dynamics CRM email management that I would like to discuss in this lesson:

> *It is very important that you never add an email address for a Dynamics CRM user to any record that is not the actual user's record.*

Many companies have employees listed as Contacts, in addition to being Dynamics CRM Users. This is not a bad practice as long as the email address on the Contact is the User's personal email address and not their corporate address.

On the outside, this does not seem like having the user's email in other parts of the system would cause a problem, but it can cause a variety of issues related to the tracking of email, whether using the Dynamics CRM Client for Outlook or the Email Router.

Issue #1: Unresolved Senders

Depending on the version of Dynamics CRM you are using, you may find that inbound email may have unresolved senders, even though they are valid users. This can be caused by the resolving mechanism locating both a User and another record, like Contact, with the same email address. Since it cannot determine which record to associate the email with, it marks the sender as unresolved.

This process has gotten better as Dynamics CRM has matured and it does not happen as much as it once did, so in your installation, it may not be an issue at all.

We covered this issue in
Email Management, so you may wish to review that lesson again if you have questions.

Issue #2: Magical Tracking of Email

Have you ever seen email within Dynamics CRM that has no business being there? Emails that are strictly internal and have no connection to any records found within Dynamics CRM?

That does happen on occasion and it is a very lengthy process to track down the circumstances of the occurrence. This may help:

> *I found out recently that any text field with the type of email within Dynamics CRM is queried by the email router when adding email to Dynamics CRM.*

This means that if you have the user's email address *anywhere* within Dynamics CRM, even inside a custom field, it can be considered as a candidate to use to set the regarding for the email.

And that is how emails get tracked accidentally.

Correcting these Issues

It is actually quite easy to correct these issues: All you have to do is remove any email address for a Dynamics CRM User that is outside of the User record. Quite simple in concept, a lot of manual labor in practice. Here are some steps that will help.

Locate Emails in Standard Entities

The Lead, Contact, and Account entities each have three email fields. Perform an Advanced Find to locate these records using the criteria seen in Figure 10-1.

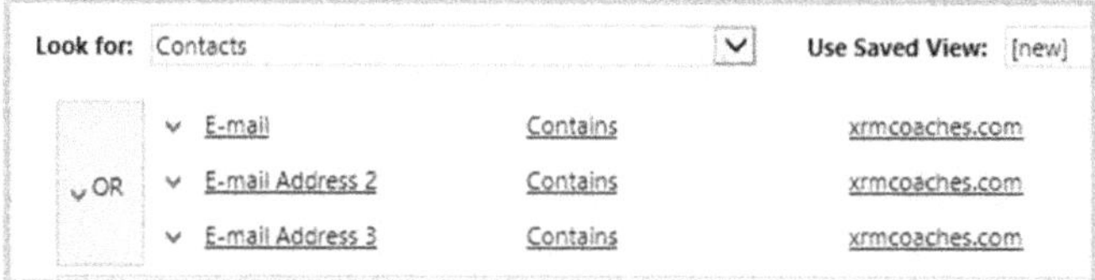

Figure 10-1. Search for a company email address

If you have more than one company domain, then you will need to modify the search appropriately. Also, I would not recommend searching on more than one entity at a time. Doing so may result in too much data to effect handle at one time.

> *I would save these searches as personal views and save them with your CRM Administrators team.*

Locating Custom Email Fields

As mentioned earlier, any text field that is of the type E–Mail will be used by the email router to locate a record to be used for association. This is the more complicated of the two processes because you must first identify the email fields.

Both the XrmToolbox for Dynamics CRM and SnapShot! for Dynamics CRM can produce reports showing the field types.

Once you have the fields identified, you need to perform the same type of Advanced Find search that we used before, but this time using the new field list.

Actual Cleanup

The biggest problem you will face is "why are these email addresses in these entities?" There may a genuine business reason why the email addresses where added, or maybe someone thought it was a good idea and just added them, no knowing the repercussions.

Once you have determined the cause, you can decide how to remove them.

Ongoing Maintenance

To prevent this issue from happening again, you can use policies to inform the user community that this is a bad thing and put into place procedures for periodically running the searches you created earlier.

Chapter 11. Outlook Sync Table Cleanup

If your organization is using the Dynamics CRM Client for Outlook, you will find a series of tables in the organization database whose names start with: **SyncEntry_** and **SubscriptionStatistics_**.

These tables are used to track items synchronized to the client machines of the users using the Outlook Client.

Should a user transfer to a different workstation or otherwise leave the company, the Sync tables will become orphaned and Dynamics CRM has no built–in mechanism to clean up or removed these tables once they become orphaned.

So, it is up to you to maintain these tables, removing any that are unnecessary. Luckily for us, the Microsoft Support Team created a SQL script to help with the cleanup.

Using the Script

The script is design to both drop specific tables as well as remove unnecessary records from associated tables. It has what I call a "practice mode," which will allow you to simply generate the SQL scripts and not execute them. This allows you to see what actions the script will be performing.

This is also a good thing to give your SQL DBA, if you have one, so they can verify and/or run the SQL scripts at a later time.

The default operation of the script will remove any data that is older than ninety days, but you can modify it if necessary.

Final Thoughts

This is generally not a script that needs to be run all that frequently, but it really depends on your environment, the number of users you have using the Outlook Client, etc.

Reference

You may read more about the script in this article:

http://blogs.msdn.com/b/crminthefield/archive/2012/10/03/cleaning-up-crm-sync-entry-tables.aspx

Chapter 12.
Monitoring the Windows Event Log

The Windows Application event log is the catch–all location where many Dynamics CRM errors are recorded. These include:

- An entire class of errors that may be considered important, but not critical, so they are not displayed within the Dynamics CRM user interface
- "Generic" errors that are shown with a minimal amount of information to the Dynamics CRM user
- ASP.NET errors
- Errors from the various services that are used by the Dynamics CRM platform

Event Sources

When Dynamics CRM is installed, it will also install the following new event sources:

MSCRMAsyncService

This is the service that runs all of the normal system jobs that maintain the system, plus any user–defined processes such as workflows and bulk–deletion jobs.

MSCRMAsyncService$Maintenance

This is a secondary version of the asynchronous service and is used to run maintenance jobs used to maintain the SQL database.

MSCRMCallout

Callouts where the plug–in technology used in Dynamics CRM version 3.0 and are no longer in use.

MSCRMDeletionService

Prior to Dynamics CRM 2011, when a user deleted a record it was not actually deleted at that point in time. It was merely marked for deletion and actually deleted once every twenty–four hours by the deletion service.

MSCRMDeployment

The deployment service is charged with things like creating new organizations, deployment administrators, etc.

MSCRMKeyArchiveManager

Archives previously generated CrmTicketKeys

MSCRMKeyGenerator

Generates CrmTicketKeys which are used to generate a security token, which helps make sure that the request originated from the user who made the request.

MSCRMKeyService

Uses the CrmTicketKeys created by the Key Generator to provide CRM security.

MSCRMLocatorService

Under investitation.

MSCRMMonitoringRuntime

This is a runtime component used by the Monitoring Service.

MSCRMMonitoringServerRole

This is a Dynamics CRM server role which is in charge of running the Monitoring Service.

MSCRMMonitoringService

The Monitoring Service periodically checks various Dynamics CRM components to see if they are in a running state.

MSCRMMonitoringTest

This is the results of a test performed by the Monitoring Service.

MSCRMPerfCounters

These are the performance counters that are built into the Dynamics CRM system.

MSCRMPlatform

The Dynamics CRM platform is the layer that sits just above the database and is in charge of interacting with the application and SDK and actually performing the operations requested by either.

MSCRMReporting

This service actually runs the SQL Reporting Services reports.

MSCRMReportingDataConnector

This service connects the Dynamics CRM web site to the SQL Reporting Services site.

MSCRMSandboxClient

Under investitation.

MSCRMSandboxService

Enables an isolated environment to allow for the execution of custom code, such as plug-ins and custom workflow activities. This isolated environment reduces the possibility of custom code affecting the operation of the organizations.

MSCRMSandboxWorker

This service is usually short–lived and is used to actually run a specific plug–in or custom workflow activity.

MSCRMTracing

This is the platform tracing service used to record detailed information about the internal functions of the Dynamics CRM platform.

MSCRMUnzipService

Handles the decompression of zipped files for data import. This service is installed as part of the Web Application Server role.

MSCRMVssWriter

Provides added functionality for backup and restore of Microsoft Dynamics CRM databases through the Volume Shadow Copy Service framework.

MSCRMWebService

This is the service that applications using the Dynamics CRM SDK actually communicate with.

Filtering Events

Depending on the software installed on your Dynamics CRM server, you may have many thousands of rows in your Application event log so it is helpful to filter the event log to help eliminate the "noise."

The Windows Event Viewer has the ability to filter your events based on a user–specified criteria. On the right–side of the Windows Event Viewer is the Actions pane, as you can see in Figure 12-1.

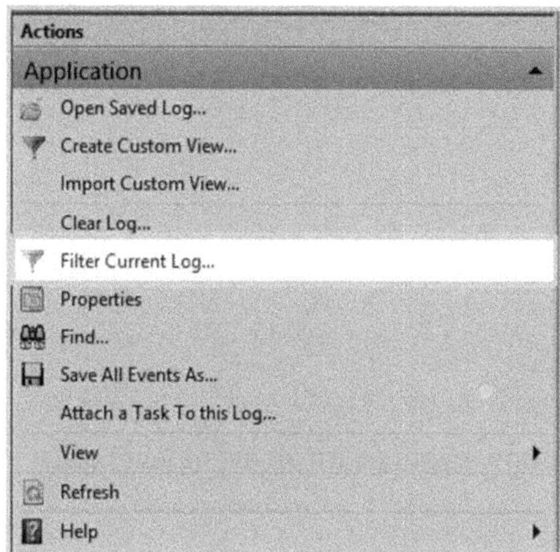

Figure 12-1. Windows Event Viewer Actions Pane

Clicking **Filter Current Log** will display the filter dialog as shown in Figure 12-2.

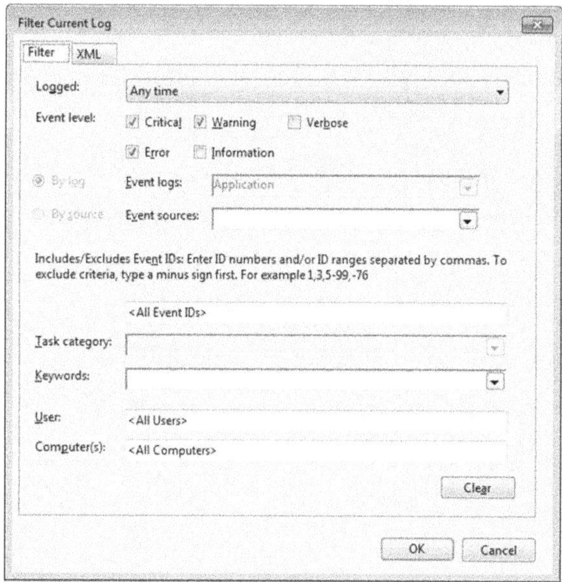

Figure 12-2. Windows Event Log Filter

Typically, you wish to filter the event log to only show events with an event level of:

- Critical
- Error
- Warning

Additionally, if you are after a very specific type of error, you can select one or more Event Sources. I usually do not set an event source because I want to see all of the errors from the entire system, instead of those from just a single, or handful, of services.

Finally, if you are searching for recent errors, you may change the **Logged** field to be:

- Last 12 hours
- Last 24 hours
- Custom range (where you supply the dates)

Searching for Issues

Now that we have our filtering in place, we can start to look for errors.

NOTES

As with the System Jobs table within Dynamics CRM, the Windows Event Log can contain a tremendous number of both informational and error messages. I have found that when troubleshooting an issue, looking at data more than 24 to 36 hours old is a waste of time. In most cases, older data is to be used to locate patterns where issues are arising based on environmental or systemic causes. The newer data should be your focus to see if you can locate the problem as it is happening or perhaps shortly after it has happened.

Remember that in many cases, if you do not catch the errors as they occur, then the probably of finding the root cause of the error after–the–fact, is rather low.

Most of the time you will find both Warnings and Errors in the Application event log. Some of those events indicate things that happened and are indeed problems, while others are just FYI events.

There is no way that I can document all of the things that you might see in the event log, but let's look at some of the most common issues:

ASP.NET 4.0.30319.0 (Warning):

An unhandled exception has occurred.

Any time you see a warning from ASP.NET it is because of an unhandled exception somewhere within either Dynamics CRM itself, or a piece of custom code. While this event is classified as a Warning, in reality, this is actually an error. These type of errors may also be seen by the user but that would depend on the type of error.

This type of event contains a tremendous amount of data so let's take a look at the event detail and discuss the major sections of interest.

```
Event code: 3005
Event message: An unhandled exception has occurred.
Event time: 1/23/2015 1:50:19 PM
Event time (UTC): 1/23/2015 6:50:19 PM
Event ID: 29ce73766e40433490bc6e9f1474eba6
Event sequence: 142521
Event occurrence: 10
Event detail code: 0
Application information:
    Application domain: /LM/W3SVC/2/ROOT-1-130664345052126052
    Trust level: Full
    Application Virtual Path: /
    Application Path: C:\Program Files\Microsoft Dynamics CRM\CRMWeb\
    Machine name: CRM2
Process information:
    Process ID: 4928
    Process name: w3wp.exe
    Account name: NT AUTHORITY\NETWORK SERVICE
Exception information:
    Exception type: FaultException`1
    Exception message: opportunity With Id = 7d8b509b-c3ff-df11-856c-00155d200c09
Does Not Exist

Server stack trace:
   at
System.ServiceModel.Channels.ServiceChannel.HandleReply(ProxyOperationRuntime
operation, ProxyRpc& rpc)
   at System.ServiceModel.Channels.ServiceChannel.Call(String action,
       Boolean oneway, ProxyOperationRuntime operation, Object[] ins,
       Object[] outs, TimeSpan timeout) at
System.ServiceModel.Channels.ServiceChannelProxy.InvokeService(IMethodCallMessage
       methodCall, ProxyOperationRuntime operation)
   at System.ServiceModel.Channels.ServiceChannelProxy.Invoke(IMessage message)
Exception rethrown at [0]:
   at System.Runtime.Remoting.Proxies.RealProxy.HandleReturnMessage(IMessage
reqMsg,
       IMessage retMsg)
   at System.Runtime.Remoting.Proxies.RealProxy.PrivateInvoke(MessageData&
msgData, Int32 type)
   at Microsoft.Xrm.Sdk.IOrganizationService.Retrieve(String entityName, Guid id,
```

```
       ColumnSet columnSet)
   at Microsoft.Xrm.Sdk.Client.OrganizationServiceProxy.RetrieveCore(String
entityName, Guid id,
       ColumnSet columnSet)
   at Microsoft.Xrm.Sdk.Client.OrganizationServiceProxy.Retrieve(String
entityName, Guid id,
       ColumnSet columnSet)
   at
Microsoft.Xrm.Client.Services.OrganizationService.<>c__DisplayClass10.<Retrieve>b
__f(
       IOrganizationService s)
   at Microsoft.Xrm.Client.Services.OrganizationService.InnerOrganizationService.
       UsingService[TResult](Func`2 action)
   at CustomPages.AccountSummaryPage.Page_Load(Object sender, EventArgs e)
   at System.Web.UI.Control.LoadRecursive()
   at System.Web.UI.Page.ProcessRequestMain(Boolean
includeStagesBeforeAsyncPoint,
       Boolean includeStagesAfterAsyncPoint)



Request information:
    Request URL:
    http://crm2:5555/myorg/ISV/CustomPages/AccountSummary.aspx?orglcid=1033&orgna
    me=myorg&type=3
    &typename=opportunity&userlcid=1033
    Request path: /myorg/ISV/CustomPages/AccountSummary.aspx
    User host address: 192.168.33.113
    User: DOMAIN\user1
    Is authenticated: True
    Authentication Type: Negotiate
    Thread account name: NT AUTHORITY\NETWORK SERVICE

Thread information:
    Thread ID: 37
    Thread account name: NT AUTHORITY\NETWORK SERVICE
    Is impersonating: False
    Stack trace:
   at System.Runtime.Remoting.Proxies.RealProxy.HandleReturnMessage(IMessage
reqMsg,
       IMessage retMsg)
   at System.Runtime.Remoting.Proxies.RealProxy.PrivateInvoke(MessageData&
msgData, Int32 type)
   at Microsoft.Xrm.Sdk.IOrganizationService.Retrieve(String entityName, Guid id,
       ColumnSet columnSet)
   at Microsoft.Xrm.Sdk.Client.OrganizationServiceProxy.RetrieveCore(String
entityName,
       Guid id, ColumnSet columnSet)
   at Microsoft.Xrm.Sdk.Client.OrganizationServiceProxy.Retrieve(String
entityName,
       Guid id, ColumnSet columnSet)
   at
Microsoft.Xrm.Client.Services.OrganizationService.<>c__DisplayClass10.<Retrieve>b
__f(
       IOrganizationService s)
```

```
   at Microsoft.Xrm.Client.Services.OrganizationService.InnerOrganizationService.
       UsingService[TResult](Func`2 action)
   at CustomPages.AccountSummaryPage.Page_Load(Object sender, EventArgs e)
   at System.Web.UI.Control.LoadRecursive()
   at System.Web.UI.Page.ProcessRequestMain(Boolean
includeStagesBeforeAsyncPoint,
       Boolean includeStagesAfterAsyncPoint)
```

The sections that contain the most valuable information are the following:

Event Code

This is the numeric value associated with this type of event. This can sometimes be used to locate more information on exactly what happened.

Event Message

The event message is more or less the title of the event. It may be a generic message, like the one above, or it may be specific.

Exception Type

This is the type of exception that occurred.

Exception message

The exception message is actually the real error message, as reported back to Dynamics CRM.

Server Stack Trace

The stack trace can be used by a developer to locate (hopefully), where the error occurred.

Request URL

The request URL is the full path of the web page that was in use when the exception occurred.

User

This is the name of the user who was using the application when the error occurred.

NOTES

Depending on the nature of the error, and the location where it occurred, the user will either get the specific exception or a generic error message displayed to them.

ASP.NET 4.0.30319.0 (Warning):

An unhandled exception has occurred.

```
An error occurred in the InvoiceCancelOpportunityCloseKeywords plug-in. account
With Id = 45b2c7ba-2039-e011-acb1-00155d200c09 Does Not Exist
```

In this case, a custom plug–in was searching for an Account that did not exist.

ASP.NET 4.0.30319.0 (Warning):

An unhandled exception has occurred.

```
Exception information:
    Exception type: InvalidOperationException
    Exception message: CRM Parameter Filter - Invalid parameter 'dType=1' in
Request.QueryString on page /myorg/_common/error/dlg_error.aspx
The raw request was 'GET
/myorg/_common/error/dlg_error.aspx?dType=1&hresult=0x80040217' called from
http://crm4:5555/myorg
/_common/error/errorhandler.aspx?BackUri=http[%]3a[%]2f[%]2fcrm4.[%]3a5555[%]2fmy
org[%]2ftools[%]2fsolution[%]2fedit.aspx[%]3fid[%]3d[%]257bfd140aaf-4df4-11dd-
bd17-
0019b9312238[%]257d&ErrorCode=&Parm0=[%]0d[%]0a[%]0d[%]0aError[%]20Details[%]3a[%
]20An[%]20unhandled[%]20exception[%]20occurred[%]20during[%]20the[%]20execution[%
]20of[%]20the[%]20current[%]20web[%]20request.[%]20Please[%]20review[%]20the[%]20
stack[%]20trace[%]20for[%]20more[%]20information[%]20about[%]20the[%]20error[%]20
and[%]20where[%]20it[%]20originated[%]20in[%]20the[%]20code.&RequestUri=[%]2fmyor
g[%]2f_common[%]2ferror[%]2fdlg_error.aspx[%]3fdType[%]3d1[%]26hresult[%]3d0x8004
0217&user_lcid=1033.
```

This error is a little ironic because it is actually stating that the dialog that is displaying an error to the user, has an error.

MSCRMPlatform (Error)

Site map for organization 89d88a39-4473-e011-8720-00155da5304e contains reference to entity new_entityname but that entity was not found in the metadata cache

This error is the result of what appears to be an unsuccessful removal of a solution. The entity, new_entityname was removed from the system, but there is still a reference to it on the SiteMap. To correct this error, you need to manually edit the SiteMap, find and remove this reference.

MSCRMPlatform (Warning):

Query execution time of XX seconds exceeded the threshold of 10 seconds

```
Query execution time of 137.9 seconds exceeded the threshold of 10 seconds.
Thread: 10; Database: MSCRM_CONFIG; Server:CRM4; Query: SELECT  Id  FROM
[MonitoringSettings]
```

Dynamics CRM has a built-in threshold of 10 seconds for any running SQL query. It does not stop the query from executing, but it will make note of the fact in the event log. There are generally two causes for this:

- Improper or out-of-date indexes on a specific table
- A system-level occurrence such as a backup operation or a genuine database-level problem

MSCRMSandboxService (Warning):

Lost contact with a Sandbox Client.

```
Lost contact with a Sandbox Client.
Source: Microsoft.Crm.Sandbox.HostService.exe (4136)
Sandbox Client: crm4: w3wp.exe (29832)
```

This event, or ones similar to it, are quite common and in most cases, can be ignored.

MSCRMSandboxService (Warning):

For improved security the Sandbox Host service account should not be a predefined local account

```
For improved security the Sandbox Host service account should not be a predefined
local account (LocalService/NetworkService/LocalSystem). A dedicated domain
account is recommended.
 Source: Microsoft.Crm.Sandbox.HostService.exe (2692)
 Service Account: NT AUTHORITY\NETWORK SERVICE
```

This is a recommendation which may or may not apply to your organization.

Chapter 13.
Platform Event Tracing Overview

Dynamics CRM has a built–in feature which allows you to trace the actual events that are occurring internally which is invaluable when troubleshooting errors.

There are two types of tracing:

- **Server–level**, which is set via the Windows Registry
- **Deployment–level**, which is set by using Windows PowerShell.

Using the Windows Registry

To enable tracing on a Dynamics CRM server, you need to add specific sub–keys to the following registry key:

```
HKEY_LOCAL_MACHINE\SOFTWARE\MICROSOFT\MSCRM
```

The sub–keys that need to be create are:

TraceEnabled (DWORD)

Value: A value of 0 or 1

If you use a value of 0, tracing is disabled. If you use a value of 1, tracing is enabled.

TraceDirectory (String)

Value: C:\Program Files\Microsoft Dynamics CRM\Trace

The TraceDirectory registry entry specifies the directory for the trace log files. The directory must exist, and the user who starts the Microsoft CRMAppPool must have full control over this directory. When you install Microsoft CRM, the default user is NT AUTHORITY\NETWORK SERVICE. This entry is only required for Microsoft Dynamics CRM 3.0. For later versions, the trace directory is set to the install location of the Microsoft Dynamics CRM program files, C:\Program Files\Microsoft Dynamics CRM\Trace

TraceRefresh (DWORD)

Value: A number between zero and 99

When the data is changed, the trace settings in the other trace registry entries are applied.

TraceCategories (String or Multi-String)

Value: Category.Feature:TraceLevel

The TraceCategories registry entry is a combination of a category, a feature, and a trace level. You can specify multiple categories, features, and trace levels. Separate each combination by using a semicolon. For a list of categories, features, and trace levels and for sample combinations that are valid, see the **Trace Categories** section for more information.

TraceCallStack (DWORD)

Value: A value of 0 or 1

If you use a value of 0, the call stack is not included in the trace file. If you use a value of 1, the call stack is included in the trace file.

TraceFileSizeLimit (DWORD)

Value: A size between 1 and 100 MB

The TraceFileSizeLimit registry entry specifies the maximum size of trace files. New files are created when the limit is reached.

Default Values

Default values for these registry sub–keys are:

- TraceCallStack True
- TraceCategories *:Error
- TraceDirectory c:\crmdrop\logs
- TraceEnabled False
- TraceFileSizeLimit 10

PowerShell

When tracing is activated using PowerShell, it is applied to the entire deployment. This is very useful if you have more than one CRM server.

```
Add-PSSnapin Microsoft.Crm.PowerShell

$Setting = Get-CrmSetting TraceSettings

$Setting.Enabled = $True
$Setting.CallStack = $True
$Setting.Categories = "*:Verbose"
$Setting.Directory = "c:\CRMtrace"

Set-CrmSetting $setting
```

Parameters

CallStack

Records callstack information. For detailed troubleshooting, we recommend that you turn this on. By default, this is turned off.

Categories

Indicates the level of detail to record.

Directory

Specifies the location of the trace log file. By default, the location is c:\crmdrop\logs.

FileSize

Specifies the maximum file size of the log file in megabytes before information in the trace file is overwritten

Default Values

The default values for these settings are:

- CallStack True
- Categories *:Error
- Directory c:\crmdrop\logs
- Enabled False
- FileSize 10

Trace Categories

ADUtility
Application
Application.Outlook
DataMigration
Deployment
Deployment.Provisioning
Deployment.Sdk
Exception
Etm
Live
Live.AggregationDataExport
Live.PartnerInteraction
Live.Platform
Live.Portal
Live.Provisioning
Live.Support
Live.SyncDaemon
Monitoring
NewOrgUtility
ObjectModel
ParameterFilter
Platform
Platform.Async
Platform.ImportExportPublish
Platform.Import
Platform.Metadata
Platform.Sdk
Platform.Soap
Platform.Sql
Platform.Workflow
Reports
Sandbox
Sandbox.AssemblyCache
Sandbox.LoadBalancer
Sandbox.CallReturn
Sandbox.EnterExit
Sandbox.StartStop
Sandbox.Performance
Sandbox.Monitoring
SchedulingEngine
ServiceBus
Shared
SharePointCollaboration
Solutions
Unmanaged.Outlook
Unmanaged.Platform
Unmanaged.Sql
Visualizations

Trace level values

Here are the valid values for the TraceLevel:

- Off
- Error
- Warning
- Info
- Verbose

Note A message is logged only if the trace level for the category is equal to or greater than the level of the message. For example, a trace level of Warning logs messages that have a level of Warning and of Error. A trace level of Info logs messages that have a level of Info, of Warning, and of Error. A trace level of Verbose logs all messages. You should use a trace level of Verbose only for short durations.

Sample category and trace level combinations:

*:Verbose

Note The *:Verbose combination logs all messages in all categories. You should only use the *:Verbose combination for short durations.

Application.*:Error

Note The Application.*:Error combination logs all messages that have a level of Error for the Application.* category.

Platform.*:Warning

Note The Platform.*:Warning combination logs all messages that have a level of Warning or Error for the Platform.* category.

this page intentionally left blank

Chapter 14.
Enabling Event Tracing

Event tracing may be enabled at the server or system level, depending on the problem you are facing, the number of servers configured, and the tool used to enable tracing.

Using CRM Tracing: Diagnostics Tool

The Diagnostics Tool will enable tracing for a single Dynamics CRM web server. The tool is actually loaded and run from the web server itself.

When you are run the Diagnostics Tool, you will see an interface like that in Figure 14-1.

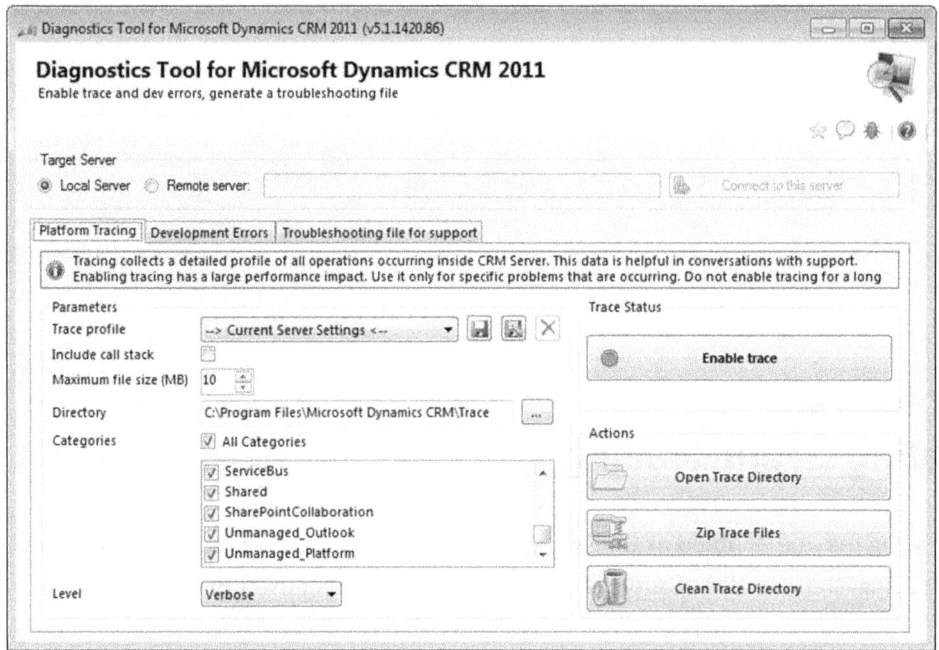

Figure 14-1. Diagnostics Tool for Dynamics CRM

> *Note: Do not let the "CRM 2011" description worry you. The process should work exactly the same on Dynamics CRM 2011, 2013, or 2015.*

Here is a review the interface components:

Target Server

Allows you to enable tracing on the local server, or on a remote server.

Trace Profile

The Diagnostic Tool allows you to create profiles which are a collection of settings. You can save these as a profile then later retrieve them. This allows you to pre–define specific types of platform tracing when attempting to diagnose specific types of problems.

Include call stack

When this option is enabled the .NET call stack is also written to the trace file. If you are attempting to locate an error in custom code, then this setting may prove useful. But, if you are just diagnosing an issue within Dynamics CRM itself, then the information may prove less than useful and can actually make reading the trace file harder.

Maximum file size (MB)

This setting controls the maximum size, in megabytes, that a trace fill will grow. When this limit is reached, a new file is created.

Directory

This is the location where the trace files will be written. This can be any folder but the service running the CRMAppPool needs access to create and write files.

Categories

These are the trace categories that you are monitoring. These will depend on the problem you are tracing. If you do not know, check the **All Categories** checkbox.

Level

This is the trace level that you would like to record. The default is Error, but in most cases, you will need to change this to Verbose.

Enable Trace/Disable Trace

This button physically enables and disables tracing on the server. You simply set the parameters as required then click the Enable Tracing button. At that point tracing is live.

Open Trace Directory

This button will open a Windows Explorer showing the Trace Directory

Zip Trace Files

This button will create a zip archive of all of the trace files in the trace directory.

Clean Trace Directory

This button will delete all of the trace files in the trace directory.

Chapter 15.
Troubleshooting Using Event Tracing

Troubleshooting using event tracing can be just a bit of a black art because of the nature of application exceptions. Here are the general steps:

1. Prepare to recreate the steps that caused the exception.
2. Turn on platform tracing
3. Recreate the exception
4. Turn off platform tracing
5. Review the trace logs to locate the error.

Trace Files

There are actually several trace files, each created by a different process.
Here is a typical trace file name:

CRM-w3wp(29832#41B3E5B8)-CRMWeb-20150125-1.log

- **CRM** is the name of the server
- **W3wp** is the name of the process
- **(29832#41B3E5B8)** is the process identifier
- **CRMWeb** is the web site
- **20150125** is the date the log was created
- **-1** is the sequential number of log file, for that day. This number will increase by one once the file size limit is reached.

Here are some of the common trace names

- CRM-w3wp is the CRM web site
- CRM-CrmAsyncService(Maintenance is the asynchronous service that handles internal maintenance jobs
- CRM-CrmAsyncService(Server is the asynchronous service that runs normal system jobs and workflows

- CRM-Microsoft.Crm.Sandbox.HostService is the sandbox service that is used to run custom code in the sandbox

It is important to understand these files because the diagnostic information recorded will be dependent on the type of issue you are troubleshooting. Most of the time the trace log you will be using will be the w3wp logs, but when you get into troubleshooting custom code, you may find yourself looking at the asynchronous or sandbox logs as well.

Viewing the Trace Logs

Once you have reproduced the situation where you encountered the error, the next step will be to view the trace log and attempt to determine the cause. For this exercise we will be using the PFE Trace Log Tool, which is a simple executable that you may install on your workstation or on the CRM server itself.

Run the application, then select File, Open, to open a trace log for viewing. The interface will resemble the one in Figure 15-1.

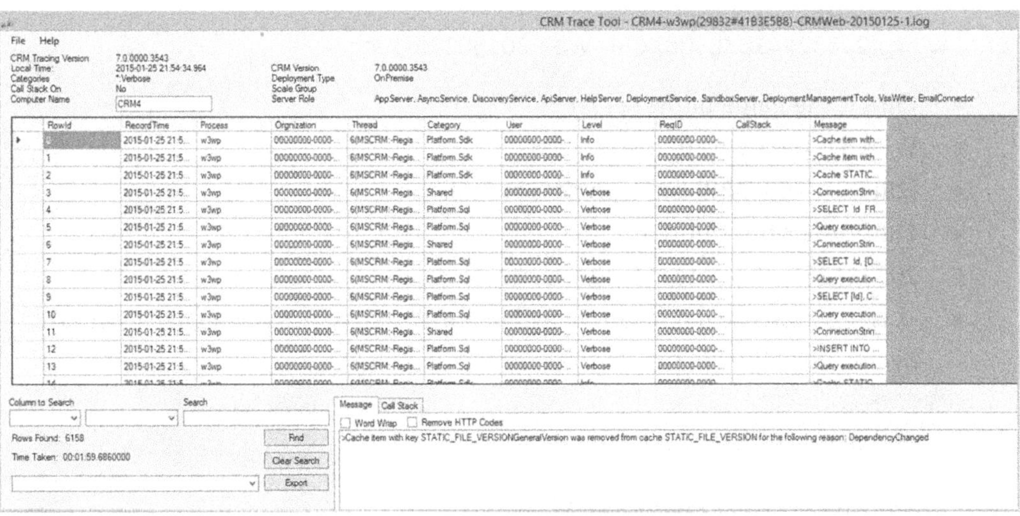

Figure 15-1. PFE CRM Trace Tool

Let's take a tour around the interface.

Header

Contains information about the server, the Dynamics CRM installation, and the trace file itself.

Grid

The grid contains the individual log entries.

Search Criteria

The lower–left corner contains a set of search criteria that allows you to search the log file for specific or general information, usually based on a specific column.

Details

The lower–right corner contains details of the log message along with the call stack, should there be one.

Locating Errors

The simplest method to locate and error is to actually search for a log entry with an Error log level. As you can see in Figure 15-2, define the following search criteria:

- Column to Search Level
- Operator =
- Search Error

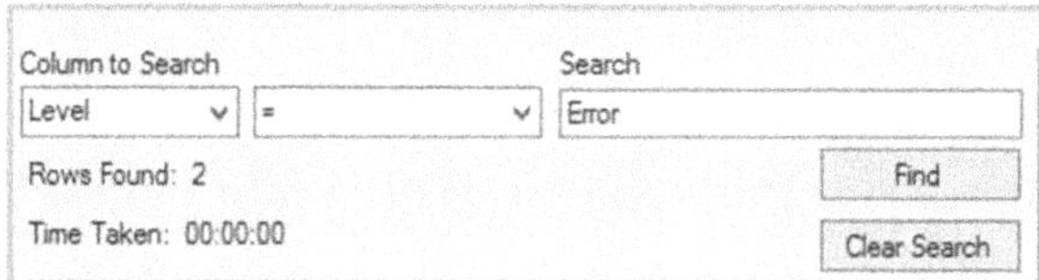

Figure 15-2. Search trace log for errors

When you click the Find button, you will see the log entries filtered to only those with a level of Error, as you can see in Figure 15-3, errors are highlighted in red. In the Message field, you can see the actual cause of the error.

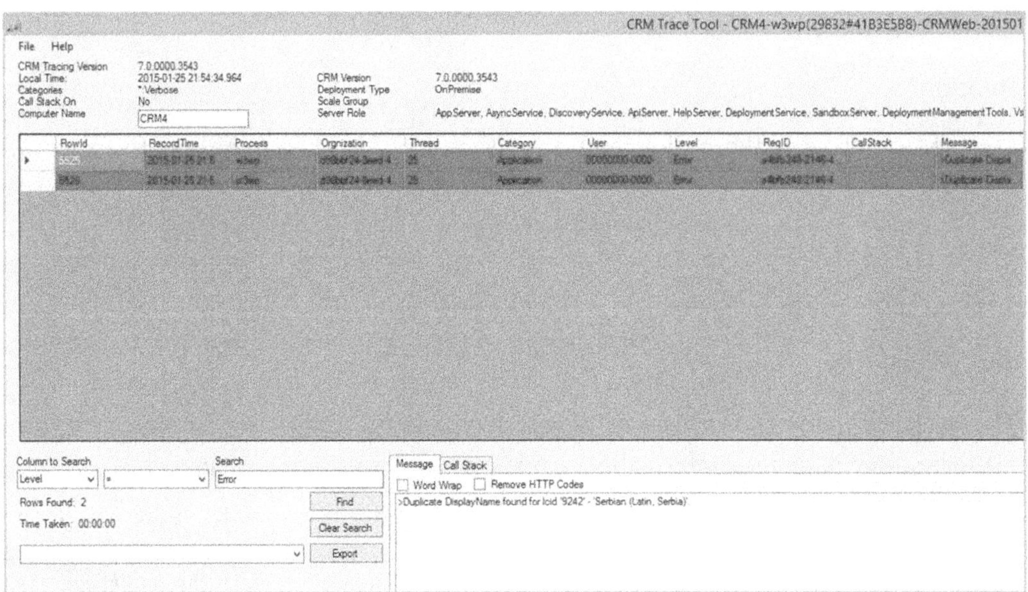

Figure 15-3. Errors found in the trace log

Most of the time, but not always, the Message will actually show you the cause of the error that was seen within the user interface.

Sometimes you will need to actually scroll through the log entries until you find the error row. Then, look at the entries above it. Sometimes errors will cause other errors and the actual root cause is not the last error in the list.

Chapter 16.
Matching User Error Codes

Depending on the situation, the Dynamics CRM application may display an error message containing an error code that may be used to locate the error. The message will look something like this:

```
Reference number for administrators or support: #7A49F455
```

The number following the # sign is not actually an error code, as you would expect, but rather it is an identifier that can be used to locate the error within the trace log file.

The only problem with this feature is that you must have tracing enabled in order for you to locate the error. The error matching code is unique and will be different every time.

If the error appears one time, then the likelihood of you finding the problem in the trace logs is rather small. If the error is repeatable, then there is almost certainly a chance that you will find it.

How the Process Works

To find an error code in the trace logs, you need to perform the following steps:

1. Move or delete any existing trace logs that are not in use.
2. Turn on tracing.

   To capture the best and most information, set the TraceCategory to *:Verbose.
3. Reproduce the error.
4. Capture the reference number shown in the error.
5. Turn off tracing.
6. Search the trace log or logs for the reference number.

   If you have a single trace file it is easiest to just open that file in the trace file viewer or Notepad++ and search for the code. If you have multiple trace files, I use AstroGrep to actually find the file with the reference number, then use the trace viewer to find the specific error.

this page intentionally left blank

Chapter 17. Troubleshooting Development Errors

If you have custom code within your environment you may, at times, encounter unexpected errors that are located in your custom code. These may or may not be shown to the user, depending on how and where the error occurred.

Usually these errors generate an error message with an associated error code. Here is an example:

```
[[Microsoft.Xrm.Sdk.OrganizationServiceFault, Microsoft.Xrm.Sdk,
Version=5.0.0.0, Culture=neutral,
PublicKeyToken=31bf3856ad364e35]]:
An unexpected error occurred.Detail:
<OrganizationServiceFault
xmlns:i="http://www.w3.org/2001/XMLSchema-instance"
xmlns="http://schemas.microsoft.com/xrm/2011/Contracts">
  <ErrorCode>-2147220891</ErrorCode>
  <ErrorDetails
    xmlns:d2p1="http://schemas.datacontract.org/2004/07/
                System.Collections.Generic">
    <KeyValuePairOfstringanyType>
      <d2p1:key>OperationStatus</d2p1:key>
      <d2p1:value xmlns:d4p1="http://www.w3.org/2001/XMLSchema"
                i:type="d4p1:int">0</d2p1:value>
    </KeyValuePairOfstringanyType>
  </ErrorDetails>
  <Message>An unexpected error occurred.</Message>
  <Timestamp>2012-11-19T15:55:39.8208026Z</Timestamp>
```

As you can see, the error message itself is not exactly helpful, so that leaves the ErrorCode as our only hope for figuring out what is happening. Here is how you locate the meaning of that ErrorCode:

1. Open the Windows calculator.
2. Set the View to Programmer.

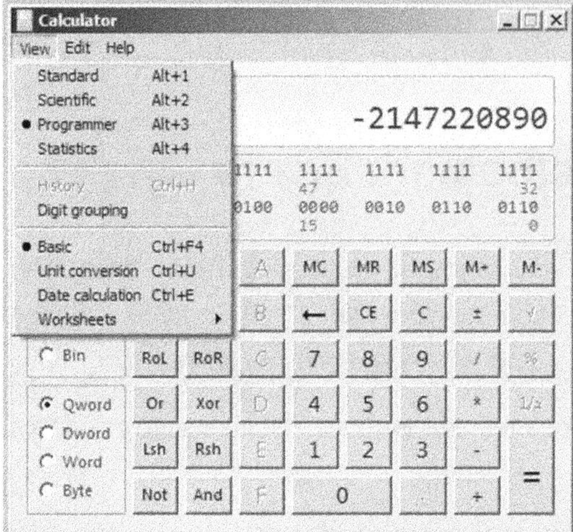

1. With the mode set to Dec, for Decimal, paste in the ErrorCode found on this line:

   <ErrorCode>-2147220891</ErrorCode>

2. Change the mode to Hex. This will convert the number from Decimal to Hexadecimal.

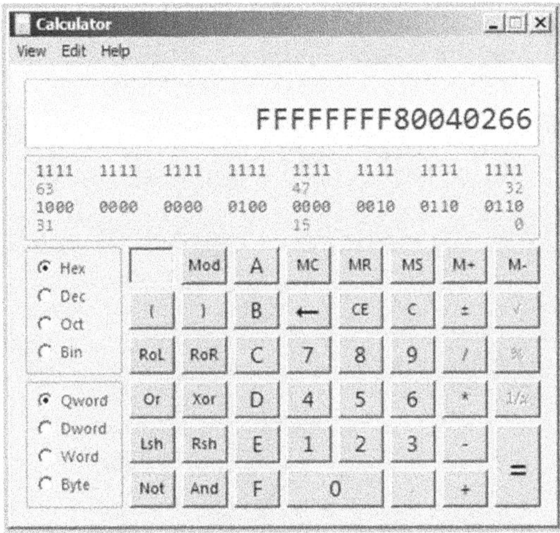

1. Select Edit, Copy to copy this value to the clipboard.
2. Open the Dynamics CRM SDK help file.
3. Paste the value you copied in Step 5 into the Search box.
4. Remove FFFFFFFF from the value.
5. Click the **List Topics** button.
6. If that error exists within the SDK, it will probably be found in the topic: *Web Service Error Codes*.
7. Click once anywhere within the topic window (on the right–side).
8. Press Ctrl+F, for Find.
9. Paste the value from the **Topic Search** box into the **Find** box.
10. You should be taken to the error code:

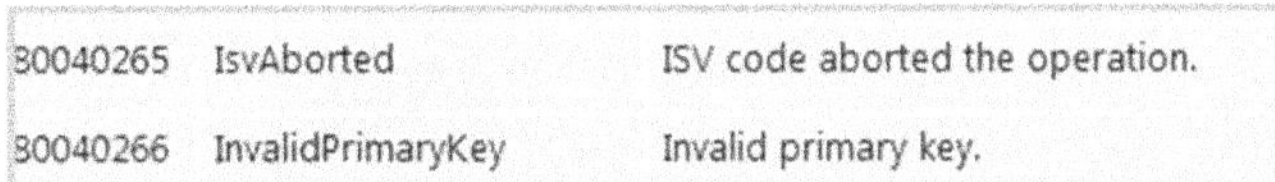

80040265	IsvAborted	ISV code aborted the operation.
80040266	InvalidPrimaryKey	Invalid primary key.

Not all error codes are documented, but most are, and this technique can really save you time when troubleshooting your code.

It should also be noted that in later versions of the Dynamics CRM SDK, both the hexadecimal and decimal codes are listed, so this technique is not necessary.

Chapter 18.
Workflow Best Practices

Review your workflows to make sure that each is needed

I hope this does not sound silly or blatantly obvious, but "feature creep" can very easily happen with workflows, as with any other part of your Dynamics CRM development process. Over a period of time, people create workflows that handle a specific task that, at the time, was very relevant to their business.

But businesses change and sometimes workflows that were valuable and necessary, become unnecessary. If you do not periodically examine your active workflows, you can end up in a situation with many workflows running on a single entity performing actions that are no longer relevant.

If you find workflows with similar functionality, consider creating a master workflow that then launches the other workflows as child workflows.

Categorize workflows by adding a keyword at the beginning of the name

If you have more workflows than will fit on a single page of the process view, it can sometimes be hard to locate a specific workflow. I found that if I add a prefix to the name of the workflow, it makes it easier to sort and group workflows by subject area. This is a very simple process, just add one of the following in front of the name of the workflow:

- Admin
- Marketing
- Sales
- Internal

Just make the list match the type and function of the workflows in your system.

Do not stop workflows with a state of Canceled

As a workflow developer you have the option of stopping the workflow at any point of the processing by using the Stop Workflow step, as you can see in Figure 18-1:

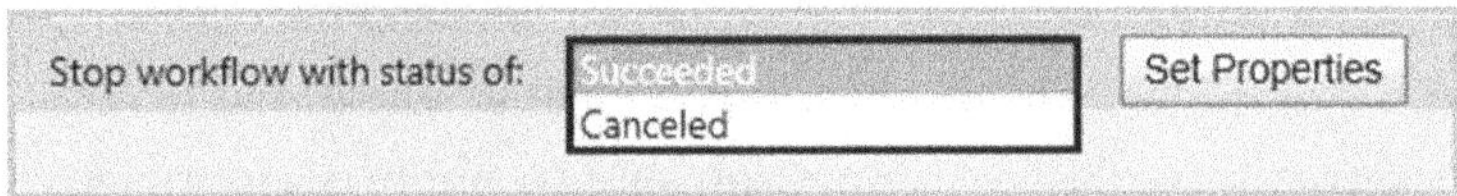

Figure 18-1. Stop Workflow Step

I think that you should always select the **Succeeded** option. If you select **Canceled**, then there is no way to determine which canceled workflows were canceled by the user or the system, and which were just terminated normally by the workflow process.

This doesn't seem like a huge problem, but it can totally invalidate your ability to troubleshoot workflow issues.

Try and be consistent with the ownership of workflows

Workflow ownership can be a little tricky, at times. You must be the owner of the workflow to activate or deactivate it. If you have multiple people performing development or administration on your system, you can end up with a variety of owners of workflows.

This can cause an issue when moving solutions containing workflows between your development, testing, and production environments.

So, if possible, it can save you a little time and trouble if you always make the owner of your workflows be the same person.

Sometimes what we will do is create a Dynamics CRM user with the System Administrator role that is used for all system maintenance, including:

- Installing solutions
- Modifying workflows
- Performing other maintenance tasks

Regularly check to ensure that workflows and system jobs are running

Simple in concept, but often overlooked: Are my system jobs being processed?

If you open the System Jobs view and see that the last date where a system job ran was more than two, maybe three, hours ago, then the probability that your Asynchronous Service is not running is very high.

The question then becomes: What do I do now?

Your first thought is probably to just turn the Asynchronous Service back on. And that could be exactly what you need to do, but it can also lead to an extreme loss of user productivity as the Asynchronous Service attempts to catch up with the backlog of jobs it has in its queue.

Depending on the configuration of your system, this can lead to the Asynchronous Service consuming a large amount of Dynamics CRM resources which can surface as a delay in the processing of user requests. This makes the system either very slow or totally unresponsive.

If this is your situation, then unfortunately, you only have two choices:

- Allow the Asynchronous Service to run through the backlog until it "catches up."
- Stop the Asynchronous Service until it can be run without impacting users (usually after–hours).

To be quite honest, neither are good options, but this is the decision that you have to make sometimes.

Monitor your asynchronous service to ensure it is not working too hard

While we are discussing the Asynchronous Service, it is very important that you monitor the amount of resources it is consuming. If it starts to consume large amounts of RAM or CPU, then you may need to take action to reduce the strain on the system.

By default, the Asynchronous Service is installed on the same server as the Dynamics CRM web site. This can lead to a fight over resources and much of the time, everyone loses.

If you do have notice an increase in resource consumption, a very quick and easy option is for you to move the Asynchronous Service to another server. Since it is just a simple Windows Service, the original service located on the web server may be disabled.

I have found that you can gain large amounts of productivity simply by moving the Asynchronous Service to another server if you have even a medium amount of system job processing to do.

Know and understand the proper use of wait conditions

The Dynamics CRM workflow engine allows you to pause the workflow to wait for a specific amount of time before continuing. Unfortunately, it is possible to use this feature incorrectly and you can end up with hundreds or thousands of workflows that will never complete.

Here is a great article that describes the proper use of the wait condition:

Workflow Wait Conditions: Best Practices - Dynamics CRM 2011

I would also recommend purchasing the book, Building Business with CRM, by Richard Knudson. This book contains a wealth of information regarding Dynamics CRM workflows and dialogs.

Draft Workflows

Draft workflows are workflows that are in the system but which are not activated and in use. Be careful of your draft workflows. If you deactivate a workflow because it is no longer used, change the name so that it is obvious that this is an unused workflow. Either that or delete it. If you have multiple administrators working with processes, it can be difficult to know which workflows are supposed to be running and which have been deactivated for a reason.

This can be especially important if you have a developer/testing/production setup where you are moving solutions from one environment to the other. Unless properly documented, the person importing the solution and activating the workflows may not know what to do.

NOTES

There are just a few of the things that I have gathered through the years. If you have additional best practices that you would like to share, then please send me an email and with your permission, I will add them to this list.

this page left intentionally blank

Chapter 19.
CRM SQL System Jobs

When Dynamics CRM is installed, it creates a series of maintenance jobs that will be run automatically at specific intervals. This is great for CRM administrators because nothing is better than a system that takes care of itself. The issue we encounter is the times when these jobs are run will start at the exact time of day when you installed Dynamics CRM.

For example, you installed Dynamics CRM in the evening, as shown in Figure 19-1. System Maintenance Job Editor, the maintenance jobs would start running at 8:45pm. That is probably ok, since most of your users may not be in the office at that time.

But what if you installed Dynamics CRM at 1:30pm. That would mean some of the most time–consuming database operations would be starting in the middle of your work day – which is not exactly good for performance.

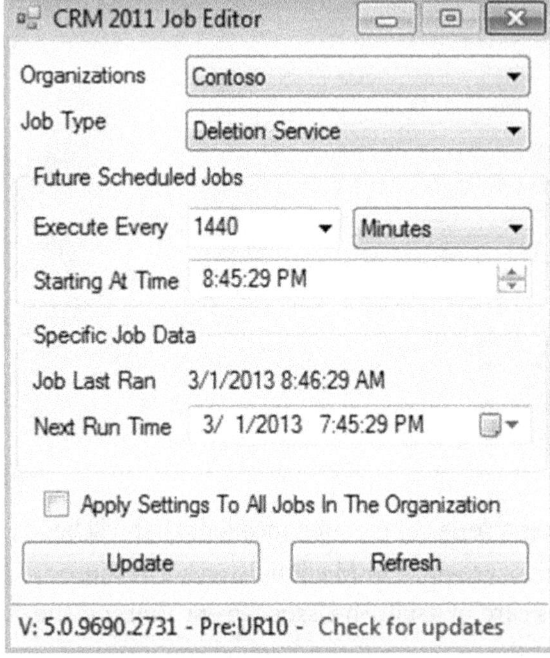

Figure 19-1. System Maintenance Job Editor

The System Maintenance Job Editor will allow you to modify the start times for these jobs. Before we get into rescheduling, let's review a few details about the maintenance jobs:

DeletionService

Frequency: Daily

In CRM 4.0, this operation had a more prominent role in that it was responsible for all physical data deletion which occurred asynchronously, but that role has been greatly diminished in 2011 since all entity data is physically deleted immediately. It now performs periodic cleanup of artifacts that were previously associated to deleted entity records such as matchcode, sync subscription, and POA records as well as the deleted object tracking records themselves.

CRM 2011 UR12: With the UR12 release, the DeletionService maintenance operation now cleans up subscription tracking records for deleted metadata objects as they expire.

CRM 2011 UR16: With the UR16 release, the DeletionService maintenance operation added a step to cleanup orphaned attachment records.

IndexingManagement

Frequency: Daily

Validates that system-managed indexes exist for all entities and recreates any missing indexes.

ReindexAll

Frequency: Daily

Reorganizes/rebuilds fragmented indexes depending on the amount of fragmentation, and performs a DBCC SHRINKDATABASE command to release unused physical drive space for both database and transaction log files. The latter works well for CRM Online organizations where drive space allocation is governed, but for on-premises environments we generally recommend postponing this job to essentially disable it (next run 1/1/2099) and opt for implementing your own index maintenance routine that does not shrink the physical files.

CRM 2011 UR12: With the UR12 release, this maintenance job has been modified to no longer shrink the database/log files as part of the operation. Thus any on-premises installation can now assess the necessity of this job based on the merits of your index maintenance strategy alone.

CleanupInactiveWorkflowAssemblies

Frequency: Daily

Seeks custom workflow assemblies that are no longer referenced in workflow rules or in-process jobs. Those unreferenced assemblies are then deleted. Consider the scenario where you register version 2.0 of a custom workflow assembly. You may update your rules to reference the new version, but some in-progress jobs may still be referencing version 1.0. Once those jobs have completed, this maintenance job will clean up the version 1.0 assembly that is no longer referenced by rules/jobs.

AuditPartitionCreation

Frequency: Monthly

Alters the partitioning scheme for the auditbase table (Microsoft SQL Enterprise only).

CheckForLanguagePackUpdates

Frequency: Daily

Detects upgrades to language (MUI) packs and schedules additional asynchronous operations to perform individual language provisioning.

RefreshRowCountSnapshots

Frequency: Daily

Refreshes the Record Count snapshot statistics leveraged in UR10's enhanced query plans.

Note: This System Job was added in Dynamics CRM 2011 UR10.

RefreshReadSharingSnapshots

Frequency: Daily

Refreshes the POA read snapshot statistics leveraged in UR10's enhanced query plans.

Note: This System Job was added in Dynamics CRM 2011 UR10.

this page left intentionally blank

Chapter 20.
CRM Organization Settings Editor

The Dynamics CRM Support team created a small solution to help you manage the settings that manage the different internal features and setting with Dynamics CRM.

The utility is actually a managed solution that is installed into a specific Dynamics CRM organization and it works with both Online and on-premises. Once installed, just open the solution and you will be taken to the solution's configuration page which will resemble Figure 20-1.

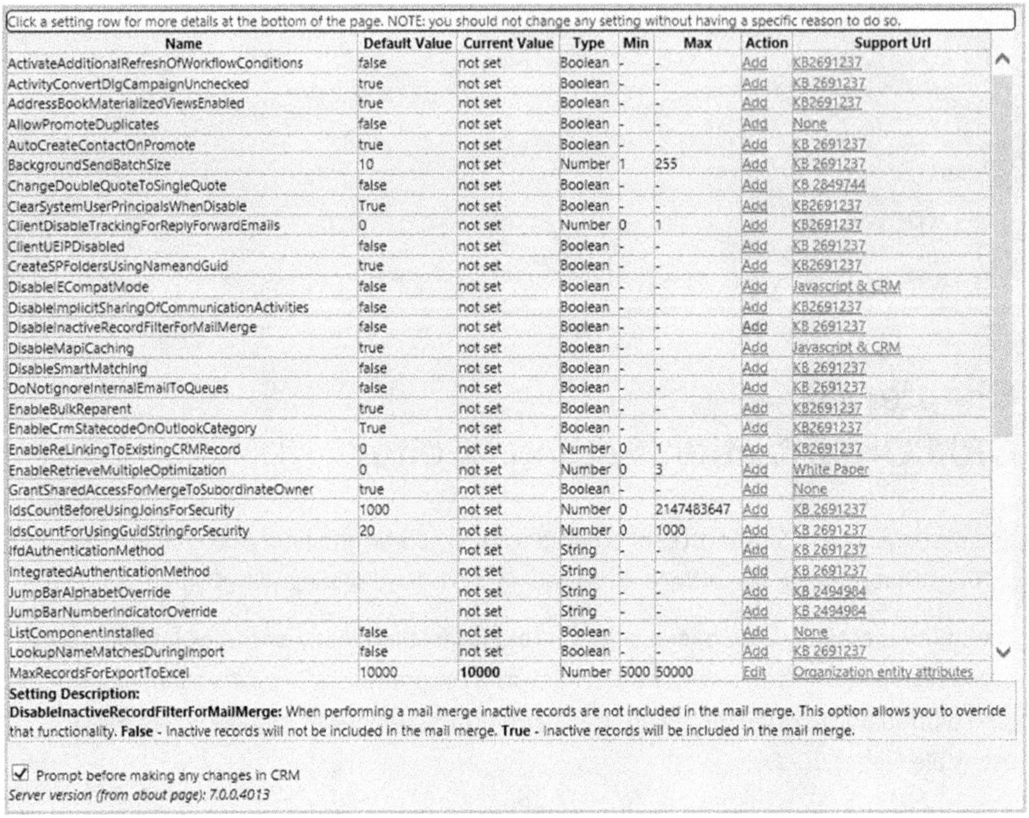

Click a setting row for more details at the bottom of the page. NOTE: you should not change any setting without having a specific reason to do so.

Name	Default Value	Current Value	Type	Min	Max	Action	Support Url
ActivateAdditionalRefreshOfWorkflowConditions	false	not set	Boolean	-	-	Add	KB2691237
ActivityConvertDlgCampaignUnchecked	true	not set	Boolean	-	-	Add	KB 2691237
AddressBookMaterializedViewsEnabled	true	not set	Boolean	-	-	Add	KB2691237
AllowPromoteDuplicates	false	not set	Boolean	-	-	Add	None
AutoCreateContactOnPromote	true	not set	Boolean	-	-	Add	KB 2691237
BackgroundSendBatchSize	10	not set	Number	1	255	Add	KB 2691237
ChangeDoubleQuoteToSingleQuote	false	not set	Boolean	-	-	Add	KB 2849744
ClearSystemUserPrincipalsWhenDisable	True	not set	Boolean	-	-	Add	KB2691237
ClientDisableTrackingForReplyForwardEmails	0	not set	Number	0	1	Add	KB2691237
ClientUEIPDisabled	false	not set	Boolean	-	-	Add	KB 2691237
CreateSPFoldersUsingNameandGuid	true	not set	Boolean	-	-	Add	KB2691237
DisableIECompatMode	false	not set	Boolean	-	-	Add	Javascript & CRM
DisableImplicitSharingOfCommunicationActivities	false	not set	Boolean	-	-	Add	KB2691237
DisableInactiveRecordFilterForMailMerge	false	not set	Boolean	-	-	Add	KB 2691237
DisableMapiCaching	true	not set	Boolean	-	-	Add	Javascript & CRM
DisableSmartMatching	false	not set	Boolean	-	-	Add	KB 2691237
DoNotIgnoreInternalEmailToQueues	false	not set	Boolean	-	-	Add	KB 2691237
EnableBulkReparent	true	not set	Boolean	-	-	Add	KB2691237
EnableCrmStatecodeOnOutlookCategory	True	not set	Boolean	-	-	Add	KB2691237
EnableReLinkingToExistingCRMRecord	0	not set	Number	0	1	Add	KB2691237
EnableRetrieveMultipleOptimization	0	not set	Number	0	3	Add	White Paper
GrantSharedAccessForMergeToSubordinateOwner	true	not set	Boolean	-	-	Add	None
IdsCountBeforeUsingJoinsForSecurity	1000	not set	Number	0	2147483647	Add	KB 2691237
IdsCountForUsingGuidStringForSecurity	20	not set	Number	0	1000	Add	KB 2691237
IfdAuthenticationMethod		not set	String	-	-	Add	KB 2691237
IntegratedAuthenticationMethod		not set	String	-	-	Add	KB 2691237
JumpBarAlphabetOverride		not set	String	-	-	Add	KB 2494984
JumpBarNumberIndicatorOverride		not set	String	-	-	Add	KB 2494984
ListComponentInstalled	false	not set	Boolean	-	-	Add	None
LookupNameMatchesDuringImport	false	not set	Boolean	-	-	Add	KB 2691237
MaxRecordsForExportToExcel	10000	**10000**	Number	5000	50000	Edit	Organization entity attributes

Setting Description:
DisableInactiveRecordFilterForMailMerge: When performing a mail merge inactive records are not included in the mail merge. This option allows you to override that functionality. **False** - Inactive records will not be included in the mail merge. **True** - Inactive records will be included in the mail merge.

☑ Prompt before making any changes in CRM

Server version (from about page): 7.0.0.4013

Figure 20-1. OrgDbSettings Settings Editor

Fields

The list of settings will vary depending on your version of Dynamics CRM. Here are the components that make up the list:

Name

This is the name of the setting.

Default Value

This is the default value.

Current Value

This is the current value. If this value is different than the default, then someone has changed it at some point in the past.

Type

This is the data type of the setting; typically: Boolean, Number, String, etc.

Min

For numeric settings, this would be the minimum possible value.

Max

For numeric settings, this would be the maximum possible value.

Action

This column allows you to either Add or Edit the current setting. If the setting does not exist within the Dynamics CRM database, then the Add action is shown. If it does exist, then the Edit action is shown.

Support Url

Since the settings list is always variable, a link to either a KnowledgeBase article or a white paper that describes the setting in detail is provided. Should no such document exist, you will see None as the value.

Editing a Setting

You may edit an individual setting when the Edit action is available. Just click Edit and the dialog shown in Figure 20-2 will appear.

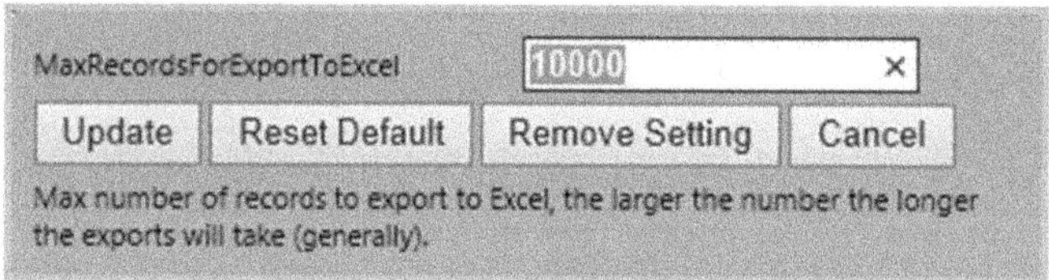

Figure 20-2. Edit an Organizational Setting

As you can see, this dialog displays the Organizational Setting, the current value, and at the bottom, it shows a description of what the setting does.

You have the following editing options:

- **Update** the value
- **Reset** the value to the default value
- **Remove** the setting from the database entirely
- **Cancel** the operation

Additional Information

In the bottom–left corner of the configuration page you will see the information shown in Figure 20-3.

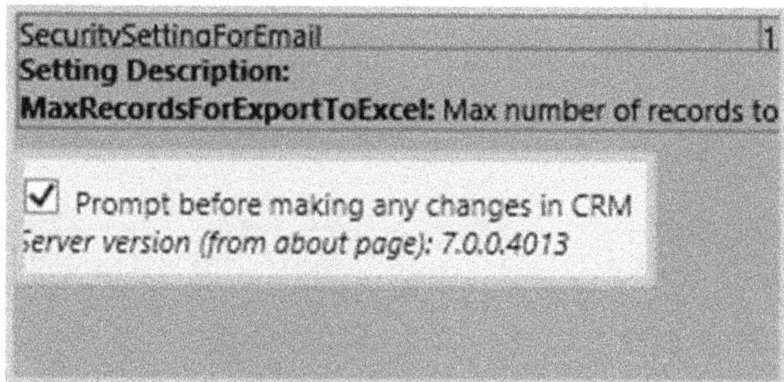

Figure 20-3. Organizational Setting Editor Options and Information

When the checkbox is checked, it will prompt you before making any changes to the Dynamics CRM database.

The line below that contains the version of Dynamics CRM you are using, as shown on the **About Dynamics CRM** page.

this page left intentionally blank

Chapter 21. Creating a Management Dashboard

Since we have created a series of saved views to help us manage the different aspects of Dynamics CRM, we can further enhance our work by creating a management dashboard.

As with the saved views, this will be a personal dashboard that you can later share with the CRM Administrators team.

Since many of the personal views we created do not have chart capability, the content of the dashboard will be simple views. Here is what I start with:

- System Jobs
- System Jobs with Messages (Recent)
- E–mails with Unresolved Senders
- Enabled Users (Detail)

This gives me a quick look at the most common views of my system. You can, of course, set these views to whatever fits your method of work.

this page left intentionally blank

Chapter 22.
Email Router Troubleshooting

Troubleshooting the Email Router can, at times, be problematic. You will often find that a system that passes the Test Access test will not actually move email properly.

Most of the time, this is related to some type of misconfiguration. Luckily for us, there is a tracing mechanism built into the email router, as with most of the platform components of Dynamics CRM.

Here is how you activate it:

1. Connect to the computer where the email router is installed.
2. Navigate to the email outer installation folder, which is typically:

   C:\Program Files\Microsoft CRM Email\Service

3. Open Microsoft.Crm.Tools.EmailAgent.xml.
4. Locate the following line:

   ```
   <LogLevel>1</LogLevel>
   ```

5. Change the value from 1 to 3.
6. Save the file.
7. Restart the Email Router service.

Unlike other Dynamics CRM platform operations, this tracing is actually written to the Windows Event Log. To view the log, perform these steps:

1. Open the Windows Event Viewer.
2. Expand Windows Logs, Application.
3. Filter the log to only show the following event levels: Critical, Error, Warning.
4. Review the filtered events for anything from the source: MSCRMEmail.

Any issues that the email router is having should appear as errors in the event log. Be warned however, there will be a massive number of informational messages from the email router which makes it difficult to actually find real errors.

Chapter 23.
SCOM Management Pack for Dynamics CRM

The Microsoft System Center Operations Manager management pack for Microsoft Dynamics CRM 2013 enables you to administer the Microsoft Dynamics CRM Server 2013 application in Microsoft System Center Operations Manager.

If you are monitoring a large number of servers, you may wish to review System Center to see if it can help your business by saving you time and money.

You may find more information about Systems Center here:

> https://technet.microsoft.com/en-us/library/hh205987.aspx

About Operations Manager

Microsoft Definition:

> *Operations Manager provides infrastructure monitoring that is flexible and cost-effective, helps ensure the predictable performance and availability of vital applications, and offers comprehensive monitoring for your datacenter and cloud, both private and public.*

Dynamics CRM Management Pack Requirements and Installation Location

You may download the Dynamics CRM Management Pack here:

Microsoft System Center Management Pack for Dynamics CRM 2013

https://www.microsoft.com/download/details.aspx?id=44279

Supported Operating System:

- Windows Server 2003, Windows Server 2008, Windows Server 2012, Windows Server 2012 Essentials

Other Software:

- System Center Operations Manager 2007 SP1 or a later version
- Microsoft Dynamics CRM Server 2013

Microsoft System Center Management Pack for Dynamics CRM 2015

https://www.microsoft.com/download/details.aspx?id=46371

Supported Operating System:

- Windows Server 2008 R2 SP1, Windows Server 2012, Windows Server 2012 Essentials

Other Software:

- System Center Operations Manager 2012 or a later version
- Microsoft Dynamics CRM Server 2015

Microsoft System Center Management Pack for Dynamics CRM 2016

https://www.microsoft.com/download/details.aspx?id=50379

Supported Operating System:

- Windows Server 2012 R2, Windows Server 2012 R2 Essentials

Other Software:

- System Center Operations Manager 2012 or a later version
- Microsoft Dynamics CRM Server 2016

Build–In Dynamics CRM Monitors

Here are the components monitored by the management pack:

Monitor	Category	Enabled	Monitored Health States
Microsoft Dynamics CRM Asynchronous Processing Service	Availability Health	Yes	Running/Not Running
Microsoft Dynamics CRM AsyncService Servers	Availability Health	Yes	Running/Not Running
Microsoft Dynamics CRM Common Class	Availability Health	Yes	Running/Not Running
Microsoft Dynamics CRM Deployment Web Service	Availability Health	Yes	Running/Not Running
Microsoft Dynamics CRM Discovery Service	Availability Health	Yes	Running/Not Running
Microsoft Dynamics CRM Email Integration Service	Availability Health	Yes	Running/Not Running
Microsoft Dynamics CRM Email Router	Availability Health	Yes	Running/Not Running
Microsoft Dynamics CRM Help Server	Availability Health	Yes	Running/Not Running
Microsoft Dynamics CRM IIS Dependent Server	Availability Health	Yes	Running/Not Running
Microsoft Dynamics CRM Organization Web Service	Availability Health	Yes	Running/Not Running
Microsoft Dynamics CRM Reporting Extensions	Availability Health	Yes	Running/Not Running
Microsoft Dynamics CRM Sandbox Processing Service	Availability Health	Yes	Running/Not Running
Microsoft Dynamics CRM Server	Availability Health	Yes	Running/Not Running
Microsoft Dynamics CRM Web Application Server	Availability Health	Yes	Running/Not Running

More information may be found in the *System Center Operations Manager Management Pack Guide for CRM 2013* document.

this page left intentionally blank

Chapter 24.
SQL Server Indexing

Creating and maintaining indexes within your Dynamics CRM database is outside of the scope of this material because it can be, and usually is, a course unto itself.

That being said, let's take a look at some of the most common scenarios where you may need to update your indexes:

Scenario #1: Windows Event Log Warning

In Chapter 12 we discussed the following error:

Query execution time of XX seconds exceeded the threshold of 10 seconds

```
Query execution time of 137.9 seconds exceeded the threshold of 10
seconds. Thread: 10; Database: MSCRM_CONFIG; Server:CRM; Query: SELECT
Id  FROM [MonitoringSettings]
```

Dynamics CRM has a built–in threshold of 10 seconds for any running SQL query. It does not stop the query from executing, but it will make note of the fact in the event log.

Again, depending on the situation, this could indicate that an index is required to improve performance. The real key to determine this is: Do you see this error repeated often, with almost exactly the same query?

If so, then you need to run that query in SQL Management Studio with the Show Execution Plan option activated so that you can see if the query can be improved by adding indexes.

Additional Information

For more information about the Execution Plan, visit the following page for a tutorial:

http://www.mssqltips.com/sqlservertutorial/285/query-execution-plans

Scenario #2: Top Queries by Average CPU Time

Microsoft SQL Server has several standard reports that you may find useful. To access them, perform these steps:

1. Open SQL Management Studio.
2. Right–click on the SQL node in the Object Explorer.
3. Select Reports, Standard Reports, Performance – Top Queries by Average CPU Time.

   This will show you a report that looks like Figure 24-1.

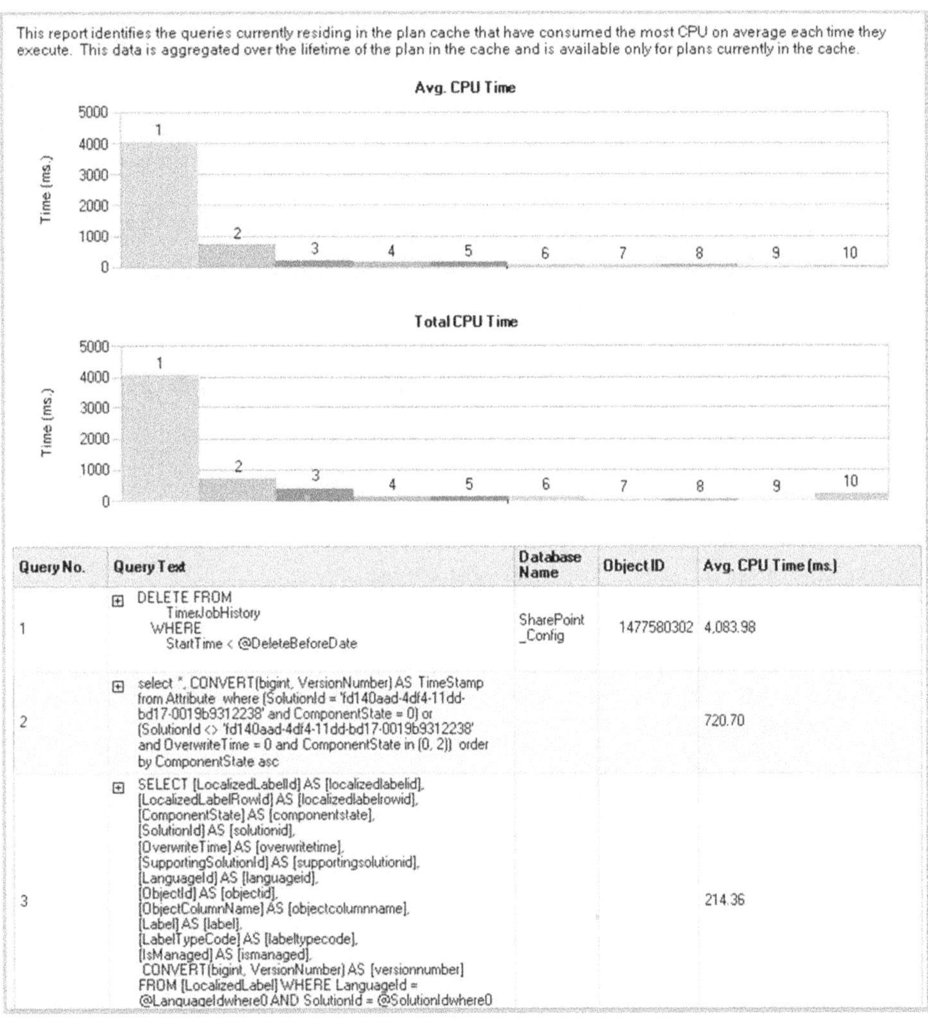

Query No.	Query Text	Database Name	Object ID	Avg. CPU Time (ms.)
1	DELETE FROM TimerJobHistory WHERE StartTime < @DeleteBeforeDate	SharePoint _Config	1477580302	4,083.98
2	select *, CONVERT(bigint, VersionNumber) AS TimeStamp from Attribute where (SolutionId = 'fd140aad-4df4-11dd-bd17-0019b9312238' and ComponentState = 0) or (SolutionId <> 'fd140aad-4df4-11dd-bd17-0019b9312238' and OverwriteTime = 0 and ComponentState in (0, 2)) order by ComponentState asc			720.70
3	SELECT [LocalizedLabelId] AS [localizedlabelid], [LocalizedLabelRowId] AS [localizedlabelrowid], [ComponentState] AS [componentstate], [SolutionId] AS [solutionid], [OverwriteTime] AS [overwritetime], [SupportingSolutionId] AS [supportingsolutionid], [LanguageId] AS [languageid], [ObjectId] AS [objectid], [ObjectColumnName] AS [objectcolumnname], [Label] AS [label], [LabelTypeCode] AS [labeltypecode], [IsManaged] AS [ismanaged], CONVERT(bigint, VersionNumber) AS [versionnumber] FROM [LocalizedLabel] WHERE LanguageId = @LanguageIdwhere0 AND SolutionId = @SolutionIdwhere0			214.36

Figure 24-1. Top Queries by Average CPU Time Report

This will show you queries from all databases on the server. You are looking for queries that are specific to your Dynamics CRM database. If you find a long–running query, you can extract that query and perform the same analysis you did in scenario #1.

Tip: The report visible within SQL Management Studio will not allow you to copy the SQL script show in the report. Export the report to a PDF file and use Adobe Acrobat to open the file. You may then copy the script from the report.

Final Thoughts on Indexing

The process of indexing your database is not a one–time event. It is something that must occur on a regular basis in order to keep the database performing optimally.

The funny thing about performance optimization, which includes indexing, is that you cannot analyze performance without usage, and you can't use the system without data.

In the case of Dynamics CRM, this means you have to actually start with the standard installation, then start tweaking performance after the system has undergone a fair amount of usage. This will allow the SQL Server to understand the database and the users using the database so it can offer up suggestions in the form of the Query Plan Optimization.

Again, without data and usage, you will not have the data you need to tune the system.

In my experience, you generally know when you need to tune when users start complaining that the system, or part of the system is slow.

It is your job to:

1. Verify the slowness actually exists at the database layer (and not custom code or the web site).
2. Locate the query or queries responsible for the slowness.
3. Tune those queries, as required.
4. Test and verify with the users that the process has worked as expected and it is now faster.

Chapter 25.
Backups, Backups, Backups

Backups are one of those things that people talk a lot about, but sometimes never do. Again, this is one of those topics we could spend days on, so it is a little outside of the scope of this course, but let's briefly discuss a few key points:

Point #1: Backup Time

Depending on the size of your database and your backup method, it may take several hours to back up the database. You need to insure that the backup process does not interfere with the productivity of either users or other system processes.

This generally means scheduling the backup during off–hours, or non–peak time.

And having a twenty four-hour operation will, of course, make that more interesting.

Point #2: Restore Time

What many people do not realize about the backup process is that the restore time is the most critical element of all.

The time to restore the backup into an operation state will be determined by how you perform the backup, or backups. Since the database and transaction logs can be backed–up separately, they must be restored separately, and in a very specific order.

The downtime realized by the organization will be totally dependent on how long it takes to recover from whatever situation that occurred. This could include:

- Restoration of the hardware
- The Dynamics CRM web site
- The SQL Server itself
- The restoration of the actual Dynamics CRM database

Again, all of these things will affect the outcome, which is to get your organization back up and running.

Point #3: Testing

It is a rather curious fact that few companies actually perform a test restore to verify that their backup process is working as desired and that their restoration process is viable.

Don't be that company.

Resources

The following resources may help you decide on a SQL backup strategy:

Best Practice recommendations for SQL Server Database Backups

http://support.microsoft.com/kb/2027537

Backup best practices – a shortlist

http://sqlbak.com/blog/backup-best-practices

SQL Server Backup Best Practices

http://sqlmag.com/database-backup-and-recovery/sql-server-backup-best-practices

SQL SERVER – Best Practices to Store the SQL Server Backups

http://blog.sqlauthority.com/2013/01/21/sql-server-best-practices-to-store-the-sql-server-backups

this page left intentionally blank

Chapter 26.
CRM Diagnostics Page

The Dynamics CRM develop team gave us a really neat little tool to help test network performance. Simply navigate to this page:

```
http://<YourCRMServerURL>/tools/diagnostics/diag.aspx
```

And you will see a page like that in Figure 26-1.

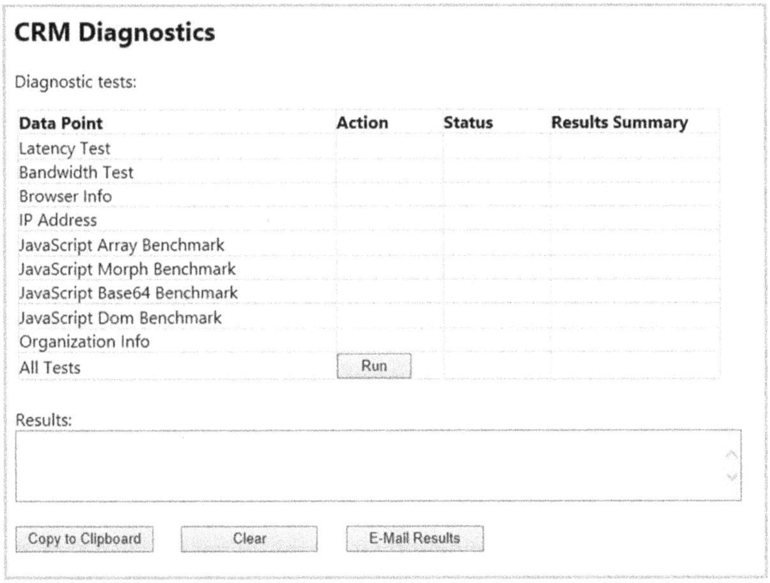

CRM Diagnostics

Diagnostic tests:

Data Point	Action	Status	Results Summary
Latency Test			
Bandwidth Test			
Browser Info			
IP Address			
JavaScript Array Benchmark			
JavaScript Morph Benchmark			
JavaScript Base64 Benchmark			
JavaScript Dom Benchmark			
Organization Info			
All Tests	Run		

Results:

Copy to Clipboard Clear E-Mail Results

Figure 26-1. CRM Diagnostics page

Operation

Once displayed, click the Run button to run the tests. The results will look something like Figure 26-2.

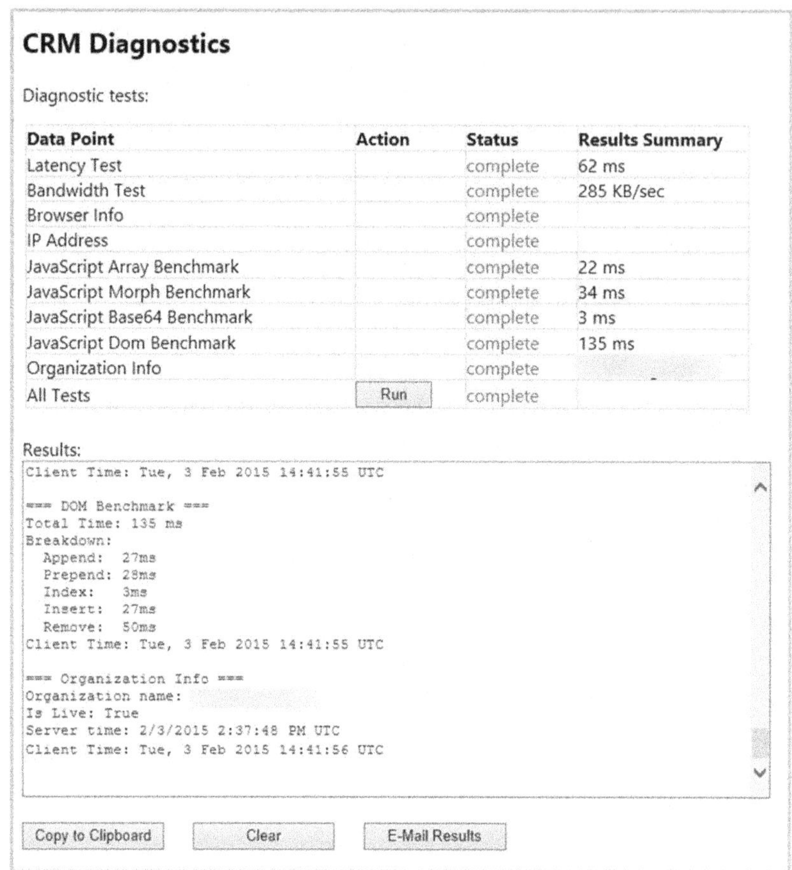

Figure 26-2. CRM Diagnostics page, post run

As you can see, it performs a variety of tests on the browser, the network, and JavaScript.

The detail from the test is in the Results field. After you have run your tests, you can copy the results to the clipboard, or email them directly from this page.

This information is useful to see if you have an issue that is related to a specific user, or the network. There may also be times, when working with the Dynamics CRM support team, that this information can be requested to help diagnose a problem – especially for Dynamics CRM Online.

Chapter 27.
Problems with Security

I cover security in-depth in my book, *Dynamics CRM Deep Dive: Security*, but we should cover the top security best practices here as well.

Don't Modify the Default Security Roles

Before you start changing security roles, make a copy of the security role that you will be changing and prepend your company's name or initials to it. Something like:

Acme-Sales Person

This will be the security role that is assigned to your sales people, instead of the default Sales Person security role.

We do this as a safety measure in case you inadvertently modify the security role to the point where Dynamics CRM no longer functions as it did before. Since you modified the copy, you can always look back at the original security role to see how it was configured out of the box.

Perform a Security Audit

If your Dynamics CRM organization has been operational for more than a few years, then it is a really good practice to perform a security audit to better understand how security has been applied. Here are the typical areas of concern:

Users with the System Administrator security role

This happens more often than you would think, but having users with System Administrator access can cause a variety of problems since they have access to the entire Dynamics CRM system. Only people who are administrators need System Administrator access so you should determine their security profile and create or assign new security roles to fit their job responsibility.

Do users have multiple security roles?

If you have multiple security roles assigned to a user that provide only slightly different access to the same entities or features, then there is maintenance issue waiting to happen.

Typically, you only have one main security role with secondary security roles that provide specific features or access to specific entities. Having for instance, the Sales Person and Sales Manager security roles create an issue because when you need to change security, you're not really sure where you should start.

Are your security roles aligned with your personnel?

Similar to the multiple security role issue, you need to ask yourself: Do your current security roles meet the needs of our users, given the jobs they perform and functionality that they require.

What features or access makes management nervous?

Often overlooked, but the concerns of management and the concerns of information technology are seldom aligned so it is important that you find out what security issues management may be concerned with when you are developing your security plan.

For instance: Deleting, Export to Excel, Assign.

This privilege is found on the Business Management tab, in the Privacy Related Privileges section, of a security role. Many companies do not like the idea of employees exporting their data to Excel and "running off with it." But, it is an extremely valuable feature used for import/export and analysis. Ask your management how they feel about this feature then make changes accordingly. Look for similar privileges as well

Bonus Material

Great! You made it this far and I want to say thank you again for your purchase!

All of the links to the tools, articles, and other relevant material may be downloaded from the following link:

http://www.xrmcoaches.com/deepdiveadministration

Please complete the form with your name and contact information and you will be emailed links to the supporting materials.

This will also add you to my mailing list and you will receive additional information as Dynamics CRM evolves and it becomes available. I purposely removed content because it was either too version–dependent or time–critical and I plan to just create additional electronic–only versions and email everyone links to the additional content.

I would also love any feedback that you may provide that will allow me to improve the content of this book.

Again, thanks for your purchase.

Sincerely,

Mitch Milam

mitch@xrmcoaches.com

www.ingramcontent.com/pod-product-compliance
Lightning Source LLC
Chambersburg PA
CBHW082035230326
41598CB00081B/6515

* 9 7 8 1 9 3 8 0 9 7 0 4 1 *